Xin Liu

Organic Semiconductor Lasers and Tailored Nanostructures for Raman Spectroscopy

Organic Semiconductor Lasers and Tailored Nanostructures for Raman Spectroscopy

by
Xin Liu

KIT Scientific Publishing

Dissertation, Karlsruher Institut für Technologie (KIT)
Fakultät für Elektrotechnik und Informationstechnik, 2015

Bildnachweis Umschlag
Prism Refraction von dgdean (Eigenes Werk)
[CC BY-SA 3.0 (http://creativecommons.org/licenses/by-sa/3.0)], via Wikimedia Commons

Impressum

KIT Scientific Publishing

Karlsruher Institut für Technologie (KIT)
KIT Scientific Publishing
Straße am Forum 2
D-76131 Karlsruhe

KIT Scientific Publishing is a registered trademark of Karlsruhe Institute of Technology. Reprint using the book cover is not allowed.

www.ksp.kit.edu

Print on Demand 2015

ISBN 978-3-7315-0370-5
DOI 10.5445/KSP/1000046874

Organic Semiconductor Lasers and Tailored Nanostructures for Raman Spectroscopy

zur Erlangung des akademischen Grades eines

DOKTOR-INGENIEURS

von der Falkutät für
Elektrotechnik und Informationstechnik
des Karlsruher Instituts für Technologie (KIT)
genehmigte

Dissertation

von

M.Sc. Xin Liu
geb. in Harbin/China

Tag der mündlichen Prüfung: 11. Februar 2015
Hauptreferent: Prof. Dr. rer. nat. Uli Lemmer
Korreferent: Prof. María A. Díaz-García

Kurzfassung

Organische Verstärkermaterialien ermöglichen durch ihre breiten Verstärkungsspektren und effiziente Energieumwandlung kostengünstige und kompakte organische Laserlichtquellen, die vielversprechend für Anwendungen in der Spektroskopie sind. Insbesondere organische Laser auf Basis der verteilten Rückkopplung (DFB, Akronym für engl. *distributed feedback*) sind leicht herzustellen und ermöglichen die Emission einzelner longitudinaler Lasermoden. Die DFB-Laseremission kann entweder in der Schichtebene oder senkrecht dazu erfolgen. Durch diese Eigenschaft werden sowohl Anwendungen als Freistrahllaserquellen als auch die Integration der organischen Laser in chipbasierten optischen Systemen ermöglicht. Zusätzlich zu ihrem Einsatz in der hochauflösenden Fluoreszenz-, Absorptions- und Transmissions-Spektroskopie, eignen sich organische Halbleiterlaser aufgrund ihrer moderaten Laserleistung und ihrer breiten Durchstimmbarkeit im gesamten sichtbaren Spektralbereich gut als Anregungsquellen in der Raman-Spektroskopie.

Diese Arbeit stellt die Anwendung von organischen Halbleiterlasern als Anregungsquelle in der Raman-Spektroskopie vor. Die verkapselten organischen Laser werden in den Aufbauten mit einem invertierten und einem aufrechten Raman-Mikroskop eingesetzt. Hierzu werden zwei Einkopplungsmöglichtkeiten untersucht: die Freistrahlkonfiguration und die Faserkopplungskonfiguration. In der Freistrahlkonfiguration wird der Laserstrahl direkt in das invertierte Mikroskop geführt. Durch Verwendung eines durchstimmbaren organischen Halbleiterlasers und mit Hilfe eines kommerziell erhältlichen Filtersystems wird eine vollständige Raman-Detektion ermöglicht. In der Faserkopplungskonfiguration wird der organische Laser durch eine Multimodenfaser in das aufrechte Raman-Mikroskop eingekoppelt. Dadurch wird ein modulares Raman-System realisiert. Um die Raman-Signale effizient zum Spektrografen zu übertragen und gleichzeitig eine ausreichend hohe spektrale Auflösung zu gewährleisten, wird ein Faserbündel eingesetzt. Die ausgeprägten Ramanspektren von Schwefel, Cadmiumsulfid und Cyclohexan werden in den experimentellen Aufbauten angeregt und detektiert. Der Vergleich dieser Daten mit Literaturwerten dient der Charakterisierung der beiden Raman-Aufbauten. Zur Verbesserung der Empfindlichkeit der Raman-Detektion wird die oberflächenverstärkte Raman-Streuung (SERS, Akronym für engl. *surface-enhanced Raman scattering*) in Experimenten eingesetzt. Der SERS-Effekt wird von goldbeschichteten Cyclo-Olefin-Copolymer (COC) Nanosäulen ermöglicht, die durch laserunterstütztes Heißprägen abgeformt werden. Die Abhängigkeit der lokalen elektromagnetischen Feldverstärkung vom Durchmesser und Abstand der Nanosäulen wird durch elektrodynamische Simulationen mittels finiter Differenzen-Methode (FDTD, Akronym für engl. *finite-difference time-domain*) untersucht. Weiterhin wird bei SERS-Messungen mit organischen Halbleiterlasern die Konzentrationsvariation des Farbstoffs Rhodamin 6G und des Biomoleküls Adenosin in wässrigen Lösungen detektiert.

Organische Halbleiterlaser können nicht nur als Freistrahl-Lichtquellen sondern auch in Lab-on-a-Chip Systeme integriert werden. Zu diesem Zweck sollte die Größe der DFB-Laser so gering wie möglich gehalten werden, um Platz für andere passive optische Komponenten zu sparen. Darüber hinaus ist es hierfür notwendig, die genaue Platzierung der DFB-Laser zu kontrollieren. In dieser Arbeit werden zwei neuartige Techniken zur Herstellung der organischen DFB-Laser in Pixelgröße untersucht: kontrolliertes Aufbringen der aktiven Materialien mit Tintenstrahldruck und präzise Herstellung der DFB-Gitter mit laserunterstütztem Heißprägen. Dieser Ansatz ermöglicht die Integration der organischen DFB-Laser in ein miniaturisiertes photonisches System für zukünftige Anwendungen in der Lab-on-a-Chip-Spektroskopie.

Abstract

Enabled by the broad spectral gain and the efficient energy conversion in the active medium, organic semiconductor lasers (OSL) are highly promising for spectroscopic applications. Especially those based on distributed feedback (DFB) resonators are simple to fabricate and are able to offer single longitudinal laser emission mode. The DFB laser emission can be out of plane of the waveguide layer as free-space laser sources or in plane as integrated laser sources in a miniaturized optical system. In addition to state-of-the-art applications in high-resolution fluorescence, absorption and transmission spectroscopy, organic semiconductor lasers are good candidates as excitation sources in Raman spectroscopy with advantages of moderate peak power and wide tunability in the visible spectral range.

This work presents the application of organic semiconductor DFB laser as free-space excitation source in Raman spectroscopy. The laser performance is improved by enhancing the laser output power and maintaining the emission stability. The encapsulated organic lasers are tested in inverted and upright Raman microscope setups, using free-beam and fiber coupling, respectively. In the free-beam configuration, the emission beam is guided directly into an inverted microscope. Employing a spectrally tunable DFB laser as the excitation source, the Raman detection is improved using a commercially available optical filter configuration. In the fiber coupling configuration, the organic laser is coupled into an upright Raman microscope through a multi-mode optical fiber with a high coupling efficiency, which provides a modular laboratory Raman system. A round-to-line fiber-bundle is utilized for an efficient collection and transfer of Raman light to a spectrograph, by keeping a sufficient spectral resolution. Raman spectra of sulfur, cadmium sulfide and cyclohexane are excited and measured in these two setups. A comparison of data from Raman characterizations and those from literature is performed to evaluate the detection systems. To improve the sensitivity of Raman detection, an approach using surface-enhanced Raman scattering (SERS) substrates is presented. Fabricated via laser-assisted replication, the SERS substrates comprise gold-coated cyclic olefin copolymer nanopillar arrays. The effect of nanopillar diameter and interpillar spacing on the local electromagnetic field enhancement is studied by finite-difference time-domain (FDTD) simulations. Furthermore, the organic-laser-excited SERS measurements are applied to verify the concentration variation of the laser dye rhodamine 6G and the biomolecule adenosine in aqueous solutions.

In addition to the use as free-space light sources, organic semiconductor DFB lasers can also be integrated into all-polymer chip platforms. For this purpose, the size of the DFB lasers is reduced as much as possible to save space for other passive optical components. Additionally, there must be an accurate lateral control to define the localized laser sites. In this work, two novel techniques for the fabrication of spatially defined organic DFB lasers are explored: local definition of the active material using ink-jet printing and laterally controlled fabrication of DFB-corrugations using laser-assisted replication. Thus this approach allows the integration of versatile coherent light sources in photonic microsystems towards lab-on-a-chip spectroscopic applications.

Contents

1 Introduction

1.1 Laser Raman spectroscopy

The advent of the first ruby laser in 1960 has set off a revolution in science and technology [1]. Due to the enormous potential of lasers in terms of temporal and spatial coherence, narrow line width, wavelength tunability and ultrashort pulse generation, the field of spectroscopy has been widened substantially since the emergence of lasers [2]. On account of the wide range of methods for laser spectroscopy, there is also a wide range of different laser sources which are used for diverse spectroscopy purposes. One of the important spectroscopic applications of lasers is Raman spectroscopy.

Since the first reports in 1928, Raman spectroscopy has been a powerful tool for the investigation of molecular vibrations and rotations in the fields of chemistry, biophysics, material science, and life science [3, 4]. It has an important advantage over infrared or near-infrared absorption for analysis of aqueous solutions, namely, practically no perturbing signal from water. In contrast to fluorescence spectroscopy, in which the molecule samples have to be labeled by a fluorescent dye to achieve a fluorescence quantum yield high enough for detection, Raman detection does not require any labeling since it is based on the intrinsic rotational and vibrational level properties of the molecule. However, the scattering cross sections in spontaneous Raman spectroscopy are very small, typically on the order of 10^{-30} cm^2, about 10^5 times weaker than the Rayleigh scattered component. In the pre-laser era, a main drawback of Raman spectroscopy was a lack of sufficiently intense radiation sources. The introduction of lasers, therefore, has indeed revolutionized this classical field of spectroscopy. Lasers have not only greatly enhanced the sensitivity of Raman spectroscopy, but they have also furthermore initiated new Raman spectroscopic techniques, such as resonance Raman spectroscopy (RRS), coherent anti-Stokes Raman scattering (CARS), hyper-Raman spectroscopy and so on.

There are some basic requirements for laser sources which must be met for Raman spectroscopy. These are: adequate laser power, high wavelength stability, high spatial mode quality and moderately narrow line width [5]. For specific applications of Raman spectroscopy, novel laser excitation sources are developed and applied in spectroscopic applications. Among them, wavelength tuning is a desired feature to conform to a specific resonance frequency and avoid fluorescence or thermal emission backgrounds. Thus, a wavelength-tunable laser is of particular interest as the excitation source for such purposes. Historically, the first tunable laser source was already organic: it was the well-known liquid dye laser. Until now, these lasers are still the most widely used types of tunable lasers in the visible and UV spectral range. However, owing to the handling complexity of toxic dye solutions and the bulky pump system, solid-state laser devices are eagerly explored. For instance, tunable infrared semiconductor-diode lasers and vibronic solid-state lasers. Meanwhile, another alternative with a wide continuous tunability over the visible range, the organic solid-state laser, demonstrates huge potential for spectroscopic applications.

1.2 Organic semiconductor lasers

Material development has played a crucial role in the investigations of new lasers. Organic lasers were discovered not long after the inorganic counterparts, but within a decade of the first laser. In 1966, the lasing effect by an optical pumping upon an organic dye solution was independently demonstrated by P. P. Sorokin and J. R. Lankard at IBM Research Laboratory [6] and F. P. Schäfer at University of Marburg [7]. The major advantage of organic active media over inorganic counterparts is the wide tunability thanks to the broad emission spectra of organic molecules. Solid-state lasers based on organic materials were demonstrated using dye-doped polymers in 1967 [8], doped single crystals in 1972 [9] and pure anthracene crystals in 1974 [10]. A milestone in the development of organic semiconductors was established in the 1970s by A. J. Heeger, A. G. MacDiarmid and H. Shirakawa. The successful synthesis and controlled doping of conjugated polymers was honored with the Nobel Prize in Chemistry in the year 2000 [11]. Since then, the applications of organic semiconductors, e.g., organic field-effect transistors (OFETs), organic light emitting diodes (OLEDs) and organic photovoltaic cells (OPVCs) have progressed rapidly. Mainly driven by the huge potential market foreseen for OLEDs as well as the progresses in both material science and deposition process techniques, optically pumped lasers based on thin films of conjugated polymers [12, 13, 14] and small organic molecules [15, 16] were demonstrated in the second half of the 1990's.

Keeping all the advantages of organic materials, optically pumped organic semiconductor lasers[1] exhibit wide and continuous tunability in the visible spectral range. Additionally, the simple and flexible fabrication processes make them free from complex handling of toxic dye solutions. Advanced lithography and deposition methods give rise to the opportunity to integrate miniaturized organic lasers of pixel-sizes into a lab-on-a-chip (LOC) platform [17, 18]. Furthermore, very low laser thresholds enable a compact optical pumping using smaller and cheaper laser diodes [19, 20, 21, 22] or light emitting diodes (LEDs) [23, 24, 25, 26]. These advantages highlight the potential of organic semiconductor lasers as excitation sources for free-space and LOC spectroscopy systems [27, 28, 29, 30]. Moreover, with advantages of moderate peak power and wide tunability, organic semiconductor lasers are good candidates as excitation sources in Raman spectroscopy.

[1] In this work, the term organic solid-state laser comprises organic semiconductor laser and solid-state dye laser.

1.3 Thesis outline

This work aims at a comprehensive exploration of organic semiconductor distributed feedback (DFB) lasers for applications in Raman spectroscopy and approaches towards spatially defined active laser sources on a miniaturized all-polymer chip platform.

The thesis is structured as follows:

In the next two chapters, the fundamental physical aspects of organic semiconductors and organic DFB lasers are described. The mechanisms of Raman scattering/spectroscopy and surface-enhanced Raman spectroscopy are elucidated. Furthermore, various fabrication techniques for organic semiconductor lasers used in this work are introduced.

Chapter 4 describes the realization of spatially defined organic lasers ("laser pixels") on an all-polymer chip platform. Two techniques, local definition of the active material and the laterally controlled fabrication of DFB-corrugations are developed. A localized manipulation of conjugated polymer via ink-jet printing and a novel fabrication of DFB-gratings via laser-assisted replication are demonstrated.

In Chapter 5, a comprehensive investigation is given to describe how to improve the performance of organic semiconductor DFB lasers for Raman spectroscopy applications. The variation of three laser components are investigated: organic active medium, DFB-corrugation substrate, and optical pumping. Subsequently, the applications of free-space organic semiconductor lasers in Raman spectroscopy are evaluated in theory and realized experimentally for the free-beam and fiber coupling configurations.

In Chapter 6, a surface-enhanced Raman scattering (SERS) approach is presented, which has the potential to significantly improve the sensitivity of Raman detection using organic semiconductor lasers as excitation sources. The SERS conditions are achieved by gold-coated cyclic olefin copolymer (COC) nanopillar arrays fabricated via laser-assisted replication. The functionality of the active SERS substrates is verified by computational simulation and experimental results. Finally, an organic semiconductor DFB laser is built and evaluated as an excitation source for SERS biomolecular analysis.

Chapter 7 summarizes this thesis and presents an outlook for further potential investigations in related topics.

2 Fundamentals

In this chapter the fundamental physical aspects of organic semiconductors: the optical and electronic properties, the gain and loss characteristics as well as the proper gain media giving rise to lasing effects are described. Furthermore various resonator configurations for building organic solid-state lasers are introduced with emphasis on one-dimensional and two-dimensional distributed feedback (DFB) approaches. The final section is given to elucidate the mechanisms of Raman scattering and surface-enhanced Raman scattering (SERS).

2.1 Organic semiconductors

Organic semiconductors are one class of organic materials, which are composed mostly of carbon and hydrogen with a few heteroatoms, such as nitrogen, oxygen or sulfur for instance. As unsaturated compounds (with double or triple bonds), organic semiconductor molecules exhibit an alternation of simple and double bonds over a planar segment. One simple constituent of this class is the benzene ring depicted in Figure 2.1(a). Three of the four valence electrons from each carbon atom form so-called σ-bonds with the hydrogen atoms and the neighboring carbon atoms through the hybridization between s- and p-orbitals as shown in Figure 2.1(b). The remaining electrons of the six carbon atoms occupy the unhybridized p_z-orbitals, which are aligned perpendicular to the plane of the σ-bonds. A weaker π-bond is formed by the overlapping p_z-electrons of two neighboring carbons (see Figure 2.1(c)).

Organic semiconductors show a network of strong σ-bonds, which define the molecular geometry, and a delocalized network of weaker π-bonds

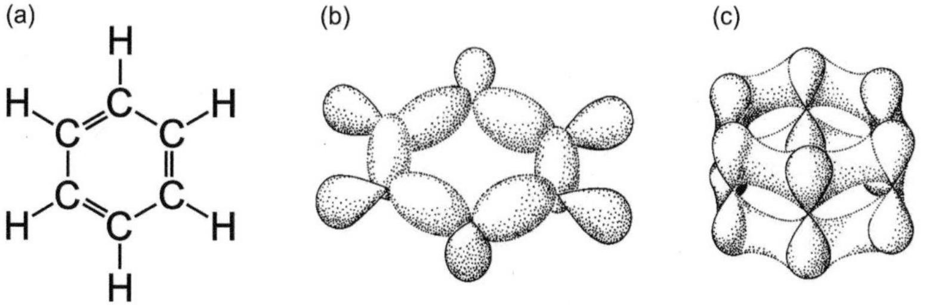

Figure 2.1: C_6H_6 (benzene): (a) chemical structure, (b) spatial distribution of the σ-bonds, and (c) spatial distribution of the π-bonds (adapted from [31]).

[31]. Two new molecular orbitals at every π-bond, the π-orbital and the π^*-orbital, emerge below and above the energy level of the separated orbitals, as shown in Figure 2.2(a). Generally, the π- and π^*-orbitals consist of discrete energy sub-levels. The electrons occupy the π-orbital before being excited to the π^*-orbital, because the energy level of the π-orbital is lower than that of the π^*-orbital. As depicted in Figure 2.2(b), benzene has six molecular orbitals, three bonding orbitals (Ψ_1, Ψ_2, Ψ_3) and three anti-bonding orbitals (Ψ_4, Ψ_5, Ψ_6). The six electrons occupy the three bonding molecular orbitals. A π-π^* transition is referred to as the transition between the ***H**ighest **O**ccupied **M**olecular **O**rbital* (HOMO) and the ***L**owest **U**noccupied **M**olecular **O**rbital* (LUMO).

2.1.1 Structural properties

Organic semiconductors are commonly classified by their chemical structural characteristics and processing techniques into two main categories: *small molecules* and *conjugated polymers*.

Different from dye lasers in liquid solution and solid-state dye lasers embedded in a non-conductive matrix, small molecule organic semiconductors are commonly fabricated into pure neat film by thermal evaporation. Since the discovery by Tang and Vanslyke in 1987 [33], the efficient electroluminescence in evaporated films of small molecules has been attracting a lot of

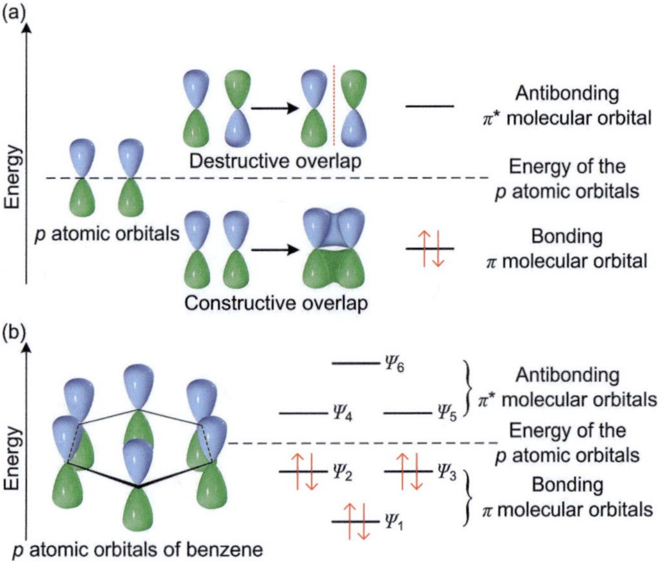

Figure 2.2: (a) The emergence of π and π^* molecular orbitals. (b) Energy level diagram of benzene for π and π^* molecular orbitals (adapted from [32]).

interest in employing these materials for light generation. Nowadays, certain small molecules are solution-processable and have already been used for the fabrication of low-cost organic optoelectronic devices [34, 35]. A typical example of small molecule materials for active laser media is the host-guest system Alq_3:DCM[1], which will be presented in detail in Section 2.2.4.

Conjugated polymers are long chain-like molecules which have alternating single and double bonds giving electron delocalization along the chain. Due to their long molecular chains, conjugated polymers are not suitable for deposition by thermal evaporation, but can be fabricated into solid-state thin films more efficiently by solution processes such as spin-coating and ink-jet printing. Two families of conjugated polymers have been widely

[1] Alq_3: Tris-(8-hydroxyquinoline)aluminum, DCM: 4-(Dicyanomethylene)-2-methyl-6-(p-dimethylaminostyryl)-4H-pyran

used for laser application: polyfluorenes [36, 37, 38, 39] (e.g., F8BT[2]) and poly(phenylene vinylene)s (e.g., MEH-PPV[3])[12, 13, 40].

Additionally, two more types of organic semiconductors which have been studied for ***Organic Semiconductor Lasers*** (OSLs): *spiro-compounds* [41, 42, 43] and *conjugated dendrimers* [44]. Spiro-compounds comprise two or more oligomer rings connected by a spiro link. They are sometimes classified into small molecule organic semiconductors. Conjugated dendrimers consist of a chromophore at the core defining the key optical properties, interior dendritic branches shielding the core and exterior functional surface groups affecting the ability and solubility.

2.1.2 Optical and electronic properties

The delocalization of the weakly bound electrons in π-bonds is responsible for the organic semiconductor properties. The energy gap between HOMO and LUMO defines the optical as well as the electronic properties of organic semiconductors [31, 45]. This work discusses mainly optically pumped organic semiconductor lasers, thus the optical properties of organic semiconductors are emphasized.

Electronic transitions and optical properties

For a better understanding of the electronic transitions in organic molecules, the π-conjugated system can be simply considered as a free electron gas enclosed in a potential well [46]. The width of the potential well is estimated according to the length of the π-conjugated system. Due to the large number of atoms in organic molecules, the energy level of the organic materials show numerous vibronic sublevels.

In virtue of the Born-Oppenheimer approximation as well as the Franck-Condon principle [47], the electrons in a molecule are supposed to move much faster than the nuclei because of their much lower mass. Using this

[2]Poly[9,9-dioctylfluorenyl-2,7-diyl)-co-1,4-benzo-{2,1-3}-thiadiazole)]

[3]Poly(2-methoxy-5-(2'-ethylhexyloxy)-1,4-phenylene vinylene)

model, the energy levels of a molecule can be determined by solving electron energies in a static molecular potential. Figure 2.3 sketches the vibrational states of each electron and their wavefunctions. The electronic transitions are depicted as vertical lines. Because the electronic transitions occur

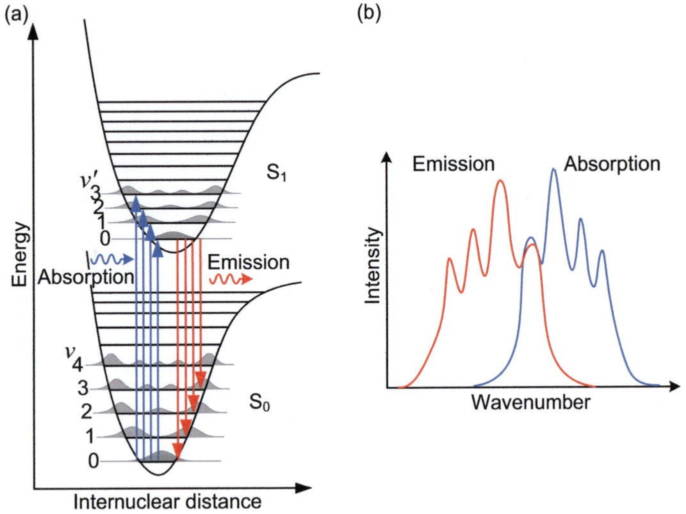

Figure 2.3: (a) Molecular potential energy as a function of the internuclear distance. The vibronic eigenstates are denoted by v and v'. The arrows indicate the vibronic transitions associated with absorption and emission of a photon. (b) The corresponding absorption and emission spectra (adapted from [48]).

in a stationary nuclear framework, the nuclear configuration of a molecule stays unchanged after absorption of a photon within a timescale of 10^{-15} s. Absorption of a photon of suitable energy causes a transition from $S_{0,v}$ to $S_{i,v'}$ (i = 1, ..., n). Afterwards the entire molecule rapidly relaxes to the lowest vibronic groundstate $S_{1,v'=0}$ by internal conversion and vibrational relaxation ($\sim 10^{-12}$ s). From there, the molecule may return to $S_{0,v}$ by a radiative transition accompanied with spontaneous or stimulated photon emission. The vibronic transition probability, given by the Franck-Condon factor, depends on the overlap of the vibrational wavefunctions of the initial and final state. The spontaneous emission transition of a fluorophore is typically a mirror image of the absorption transition, as illustrated in Figure 2.3(b). This is due to the fact that the probability of a molecule returning to a particular vibrational energy level in the ground state is governed by the same overlap factor as the absorption process [48]. The existence of a large red-shift of the emission spectrum relative to the absorption spectrum, known as *Franck-Condon shift* or *Stokes shift*, is an important property in organic active semiconductors for laser applications, as it reduces the self-absorption loss and hence reduces the laser thresholds [49].

A comprehensive overview on various transition processes in an organic molecule can be best presented with the help of a Jablonski diagram, as illustrated in Figure 2.4. It shows a simple schematic of a typical molecule with singlet and triplet states, where the singlet states are marked with S_0, S_1, ..., S_n, which have a total spin of zero, and the triplet states T_1, T_2, ..., T_n, which have a total spin of one, respectively. The lowest singlet state is S_0, which is also the lowest energy level, whereas the lowest triplet state is not T_0, but T_1 according to the Pauli exclusion principle. As previously discussed, molecules are typically excited to a vibrational level of one of its electronically excited singlet states $S_{i,v'}$ by the quick absorption ($\sim 10^{-15}$ s) of ultraviolet or visible light. Direct excitation from the ground state $S_{0,v=0}$ to an excited triplet state is highly unlikely, because the transition is spin-forbidden. Once a molecule is promoted to the S_1 state, either directly from $S_{0,v=0}$ or after *internal conversion* ($\sim 10^{-12}$ s) from a higher excited singlet state (S_2, S_3, ..., S_n), it decays to $S_{1,v'=0}$ by a rapid *vibrational relaxation* process, occurring within $\sim 10^{-14}$–10^{-10} s. Then the molecule is deactivated through different possible mechanisms. The molecule can relax radiatively either through *fluorescence*, which has a typical lifetime of $\sim 10^{-9}$–10^{-7} s or through stimulated emission. It can also decay nonradiatively directly to the ground state by *internal conversion*. Additionally, the molecule has a possibility to undergo a transition to the long-lived triplet state T_1 through *intersystem crossing* ($\sim 10^{-10}$–10^{-8} s).

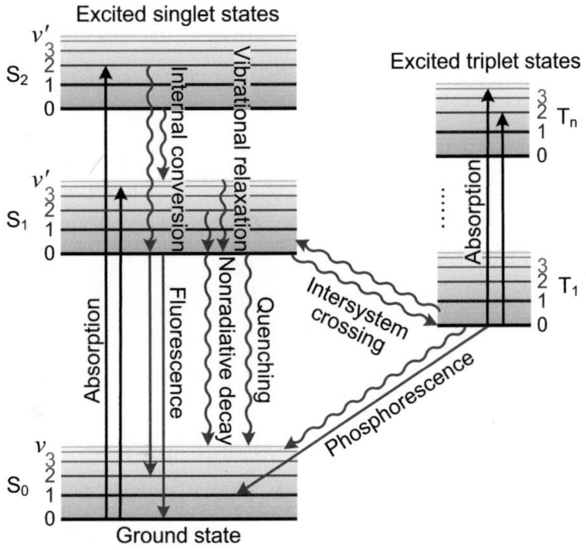

Figure 2.4: Jablonski-diagram illustrating the main photophysical processes in an organic molecule (adapted from [31]).

It results either in an emission of a photon through *phosphorescence* or a transition back to the excited singlet state, which can yield *delayed fluorescence*. Transitions from the triplet excited state to the singlet ground state are forbidden, and therefore the lifetime of phosphorescence ($\sim 10^{-3}$–10^{2} s) are several orders of magnitude longer than that of fluorescence.

Electronic properties

To study the electronic properties of the organic semiconductors, let us start from their counterparts, *inorganic crystalline semiconductors*. The covalently bound atoms in these materials are periodically arranged in a crystalline structure. This periodicity allows a delocalization of electrons, holes or coulombically bound excitons over the entire crystal. The charge carriers can be described by Bloch waves, thus leading to a high charge carrier mobility [50]. A typical electron mobility for silicon at room temperature is ~ 1400 cm^{2}V^{-1}s^{-1} and the hole mobility is ~ 450 cm^{2}V^{-1}s^{-1} [51]. Conversely, the majority of organic semiconductors for organic electroluminescent devices are amorphous. As elucidated previously, the conductivity of organic semiconductors is the result of conjugated π-systems with alternating single and double bonds. The absence of well-defined valence and conduction bands owing to the disorder of amorphous organic materials leads to a low charge carrier mobility. In the best cases, the room-temperature mobility of crystalline organic semiconductors can reach up to 0.1–20 cm^{2}V^{-1}s^{-1}. In amorphous materials, the mobility generally drops well below 0.1 cm^{2}V^{-1}s^{-1} [52]. The conjugated polymers have typically room-temperature mobilities of $\mu_e = 10^{-5}$ cm^{2}V^{-1}s^{-1} for electrons and $\mu_h = 10^{-3}$ cm^{2}V^{-1}s^{-1} for holes, respectively [49, 53].

In these highly disordered systems, the band transport is replaced by thermally activated hopping of charge carriers between localized states [47, 54, 55, 56]. Higher temperatures improve the transport by providing the energy required to overcome the barriers created by energetic disorder [52, 54]. To improve the low conductivity of organic semiconductors, defined as $\sigma = e \cdot n \cdot \mu$, where e denotes the elementary charge, n the carrier concentration and μ the charge carrier mobility, more charge carriers can be added to the organic materials. This is accomplished by charge injection and doping [57, 58, 59, 60, 61, 62, 63].

2.2 Lasing in organic semiconductors

The word LASER is the acronym for ***L**ight **A**mplification by **S**timulated **E**misson of **R**adiation*. In general, a laser consists basically of three fundamental elements: an amplifying material (active medium), a resonator (cavity) and a pump system. To realize light amplification, there must be a *population inversion* between energy levels to favor stimulated emission over absorption for a given transition [64, 65]. In optical pumping, at least three different energy levels are required to enable a population inversion.

A real laser can be characterized by the following properties: an extreme narrow line-width emission, a clear threshold in both output power and line width versus pump power and the influence of resonator geometry on the laser emission properties [66].

2.2.1 Stimulated emission

The optoelectronic properties enable the organic semiconductors to behave as four-level laser systems [67, 68, 69]. The advantage of a four-level laser system over a three-level laser system embodies in a beneficial population inversion. Figure 2.5(a) illustrates a four-level laser system. There can be a population inversion between the levels <3> and <2>, even when most of the molecules are in the ground state. Therefore the lasing can be obtained with a low threshold [64].

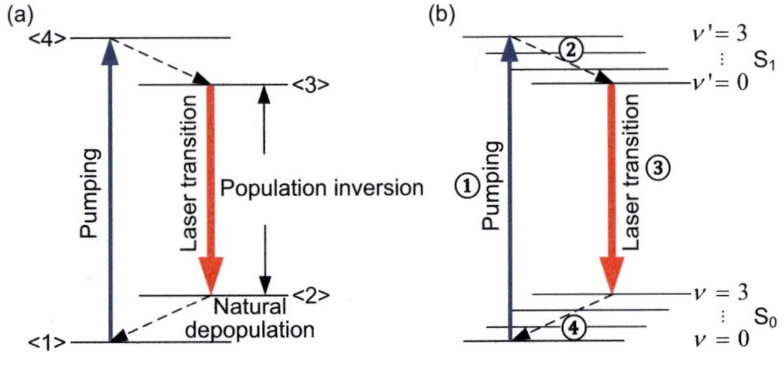

Figure 2.5: (a) Energy levels and transitions of a typical four-level laser material. (b) Energy levels of the two lowest singlet states in an organic semiconductor.

The energy levels in a typical organic semiconductors are shown in Figure 2.5(b) with ground state and the first excited singlet state. Each of the electronic energy levels is divided into vibronic sublevels. External photons can excite the molecule from its ground state to an excited vibrational level of the singlet manifold (corresponding to transition ① in Figure 2.5(b)). The molecule relaxes nonradiatively to the bottom of the singlet manifold (transition ②). Lasing can take place on transition ③ to a vibrationally excited level of the ground state manifold, followed by a vibrational relaxation (transition ④). It also explains why the emission occurs at longer wavelength than the absorption.

To dwarf the spontaneous emission, there must be enough light amplification to sustain stimulated emission, which is supplied either by a large gain in medium or low loss feedback from an efficient cavity. As light travels through an amplifying medium, its intensity grows exponentially according to

$$I = I_0 \exp[(g(\lambda) - \alpha(\lambda))L] \tag{2.1}$$

where I_0 is the intensity of incoming light, $\alpha(\lambda)$ the loss coefficient, L the distance traveled in the gain medium and $g(\lambda)$ the spectral gain coefficient. To evoke the laser oscillation, the gain must be greater than the respective losses:

$$g(\lambda) > \alpha(\lambda) \tag{2.2}$$

The necessary value for g defines the lasing threshold.

The spectral gain $g(\lambda)$ depends on the spectral cross section of stimulated emission σ_{SE} and the volume density of excited states N_{exc} [67, 70]. The net gain ($g_0(\lambda) = g(\lambda) - \alpha(\lambda)$) can be obtained using the variable stripe length (VSL) technique[4] [72, 73, 74, 75] or transient absorption measurements [76, 77, 78, 79, 80, 81, 82]. Organic active media have large gain cross sections around 10^{-17}–10^{-15} cm^2 [68, 83, 84, 85], which are much larger than those of Nd:YAG[5] lasers (around 10^{-19}–10^{-18} cm^2) [86, 87] and comparable to the gain cross sections in inorganic semiconductor laser diodes, e.g., indium gallium nitride (InGaN) [88].

[4] Basically the VSL technique can only be used to directly measure the net gain of the material. However, the waveguide propagation losses α can be determined through modification in the setup and moving the pump stripe with a constant length away from the sample edge [69, 71].

[5] neodymium-doped yttrium aluminum garnet

2.2.2 Amplified spontaneous emission

Under optical pumping (the energy transfer from an external light source into the active medium), the light may emerge from the active medium in form of amplified fluorescence, when the net gain is present in the medium. It exhibits a much narrower spectrum than the fluorescence spectrum and is often referred to as "mirrorless lasing", i.e., *amplified spontaneous emission* (ASE).

The ASE phenomena give evidence of stimulated emission in a given medium. In the previously mentioned VSL technique, ASE measurements are applied to determine the net gain and propagation loss of a certain waveguide structure. In this method, the sample is normally pumped by a beam with a tailored form of a thin stripe of variable length. The wavelength dependent output intensity of the ASE, $I(\lambda)$, is given by [72, 73]

$$I(\lambda) = \frac{A(\lambda)I_{\mathrm{p}}}{g_0(\lambda)}[e^{(g_0(\lambda)l)} - 1] \tag{2.3}$$

where $A(\lambda)$ is a constant related to the spectrally dependent emission cross section, I_{p} is the pump intensity, and l is the stripe length. By monitoring the ASE intensity as a function of the variable stripe length, one may calculate the net gain $g_0(\lambda)$.

The ASE spectra, which are typically with a *full width at half maximum* (FWHM) of $\sim$ 5–20 nm [49, 69, 73], are narrower than the fluorescence spectra, but still broader than the laser emission spectra, as represented in Figure 2.6. As there is also a competition between optical gain and loss before ASE phenomenon arises, an ASE threshold can be measured and characterized as another important parameter in evaluating the suitability of a material for lasing. Lower ASE thresholds in active media correspond to lower lasing thresholds for the laser devices based on similar resonators.

As "mirrorless lasing", ASE in waveguide structures share many properties of a real laser, such as a distinct threshold in the input-output characteristics and a polarized, concentrated and almost monochromatic beam.

However they are not lasers. The absence of resonant mode properties and the incoherent emission beam distinguish ASE from the lasing effect [66].

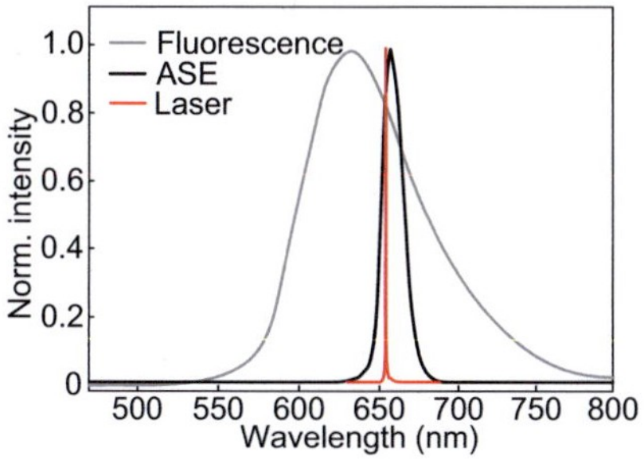

Figure 2.6: Fluorescence, ASE and laser spectrum of an organic semiconductor laser consisting of a neat film of Alq_3:DCM with a film thickness of 350 nm on a one-dimensional glass grating substrate with a corrugation period of 400 nm.

2.2.3 Loss mechanisms in organic semiconductors

Note that the net gain is the material gain minus the losses $(g(\lambda) - \alpha(\lambda))$. These losses might be due to light scattering, outcoupling out of the resonator and reabsorption at the emission wavelength. Among them, light scattering and outcoupling losses are related to the device structure comprising the resonator configuration and the organic thin film. They can be reduced by improvements in resonator quality as well as thin layer homogeneity. The reabsorption and the excited state absorption losses are associated with the photophysical fundamentals of organic semiconductors and are discussed in detail below.

As elucidated in Section 2.1.2, the molecule at the first excited singlet state S_1 has a possibility to undergo a transition to the long-lived triplet state T_1 through intersystem crossing. This may be unfavorable for stimulated emission transition taking place at $S_{1,v'=0} \rightarrow S_{0,v}$ [67, 89]. The long lifetime of the triplet state in organic active medium is on the order of milliseconds [90], which leads to a high possibility of triplet-triplet absorption. However, the $T_1 \rightarrow T_n$ absorption bands are broad and have significant overlapping with the fluorescence and laser emission spectrum, therefore this is detrimental for the laser efficiency. In addition, the recovery of $T_n \rightarrow T_1$ through a fast nonradiative decay might not be complete. The remaining excited triplet states may react with oxygen in the environment and form carbonyl bonds (C=O) or end up in electron-hole pairs [91, 92, 93]. Both, the photooxidation and polaron pairs, will cause luminescence quenching and hamper laser operation [78, 94]. In addition, the dynamic reduction of the volume density of S_1 excited state ΔN_{exc} also lowers the optical gain $g(\lambda)$ of the stimulated emission, i.e., the lasing effect will probably be suppressed.

Since the triplet-triplet absorption transition is spin-allowed, its cross section is relatively large ($\sim 10^{-19}$–10^{-16} cm^2 [95, 96, 97]) and it is comparable to the stimulated emission cross section. An elimination of the triplet states is required to obtain a better organic laser performance. To avoid triplet accumulation, the great majority of optically pumped organic semiconductor lasers are operating with femtosecond to nanosecond pulse duration, allowing the triplet population to pour out between two excitation pulses [98]. The typical maximum pulse length of an organic laser must be significantly smaller than the triplet lifetime and is expected to vary from a few picoseconds to a few microseconds depending essentially on the intersystem crossing lifetime [99, 100]. The typical optical pump repetition rate is between 10–10^4 Hz. Towards quasi-CW organic lasers, the technique has improved for certain materials up to several MHz [101, 102]. One proposal of CW laser used a rotating disc as gain medium to mimic a liquid dye laser circulation [103]. However, the fast disk rotation led to a noisy and unstable output. Until now, true CW lasing in organic semiconductors has not yet been achieved. It would only be possible either when the triplet absorption probability were lower than the singlet stimulation emission probability or a phosphorescent laser was enabled. Both breakthroughs require a small triplet-triplet absorption cross section, a short triplet lifetime in the organic active medium and a small spectral overlap between triplet-triplet absorption and fluorescence emission.

Adding a proper triplet scavenger or a triplet manager to the active media may facilitate the radiative singlet stimulated emission while suppressing triplet-triplet absorption [104], which has been experimentally achieved for organic solid-state lasers [100, 105].

In addition to the negative influence of triplets on laser action, the singlet excited states can be reduced further by bimolecular interactions, e.g., singlet-singlet annihilation (SSA), singlet-triplet annihilation (STA) [69, 98].

2.2.4 Materials for organic semiconductor lasers

In Section 2.1.1 a classification for organic semiconductors was given. In this section the main organic active media employed in this dissertation will be described and some details about their chemical structure and optical properties will be presented. The basic criteria for choosing the active medium in this work are outlined as follows:

1. A diode pumped solid-state (DPSS) UV-laser with an emission wavelength of 355 nm and a pulse duration of ~ 0.5 ns is the optical pump source for the organic lasers. The photons of this wavelength should be efficiently absorbed by the employed active medium or one of the composite material systems (e.g., organic hosts for energy transfer). It means, the absorption spectrum of the active medium should have a relatively high peak value at 355 nm.

2. A high luminescence efficiency is required.

3. The material system is stable against environmental moisture and oxygen.

4. The fabrication process is easy to handle. The fabrication methods are introduced in Chapter 3.

5. A large overlap of donor emission spectrum with acceptor absorption spectrum in guest-host systems based on Förster resonant energy transfer (a large Förster radius of several nm) is desirable.

Förster resonant energy transfer

As a chromophore molecule is surrounded by identical chromophores and impurities, such as different kind of chromophores, dissolved oxygen or water and so on, the interaction among them is unavoidable. It may be accompanied by detrimental quenching of emitting singlet states or conversely useful energy transfer. *Förster resonant energy transfer* (FRET) belongs to the latter. As first discovered by Th. Förster in 1948 [106], it is a nonradiative dipole-dipole interaction between the donor and the acceptor chromophores. This phenomenon does not require a direct contact between two chromophore molecules, but an overlap between the emission spectrum of a donor molecule (host) and the absorption spectrum of an acceptor molecule (guest). FRET is a long-range energy transfer (30–100 Å) in comparison with *Dexter energy transfer*, where donor and acceptor must be very close to each other (6–20 Å) so that simultaneous transfer of electrons and holes can happen. The FRET transfer rate k_{RET} can be expressed as [47]

$$k_{\mathrm{RET}} = \frac{1}{\tau_{\mathrm{D}}}\left(\frac{R_0}{r}\right)^6 \tag{2.4}$$

where τ_{D} is the donor lifetime and R_0 is the Förster radius, defined as the distance for which spontaneous decay of the donor and energy transfer to the acceptor are equally probable:

$$\left(\frac{R_0}{r}\right)^6 = \frac{1}{r^6}\frac{3}{4\pi}\int_0^\infty \frac{c^4}{n^4\omega^4} f_{\mathrm{D}}(\omega)\sigma_{\mathrm{A}}(\omega)\mathrm{d}\omega \tag{2.5}$$

where $f_{\mathrm{D}}(\omega)$ and $\sigma_{\mathrm{A}}(\omega)$ denote the normalized donor fluorescence spectrum and acceptor absorption spectrum respectively, n is the refractive index of the host. A large Förster radius suggests an effective transfer.

FRET occurs only between dipole-allowed transitions, i.e., singlet-singlet, not singlet-triplet [53, 106]. It is efficient if the overlap between the emission of donor and the absorption of acceptor is large.

For laser applications, one example of FRET is the use of a blend of the red-emitting laser dye DCM as a dopant in the green-emitting host material

Alq_3. The conjugated polymer lasing system $F8_{0.9}BT_{0.1}$:MEH-PPV used in this work is also based on FRET. The Förster energy transfer is employed to spectrally shift the emission away from the absorption band, which can lower the self-absorption losses of organic materials and hence achieve very low laser thresholds [107, 108].

The composite small molecule system Alq_3:DCM

The guest-host system comprising of the host coordination complex tris(8-hydroxyquinolinato)aluminium (Alq_3) and the guest laser dye 4-(Dicyano-methyl)-2-methyl-6-(4-dimethyl-amino-styryl)-4-H-pyran (DCM) is one of the first organic electroluminescent materials used as the active medium for optically pumped organic solid-state lasers [15, 107, 109]. The chemical structures of these two small molecules are shown in Figure 2.7(a) and 2.7(b).

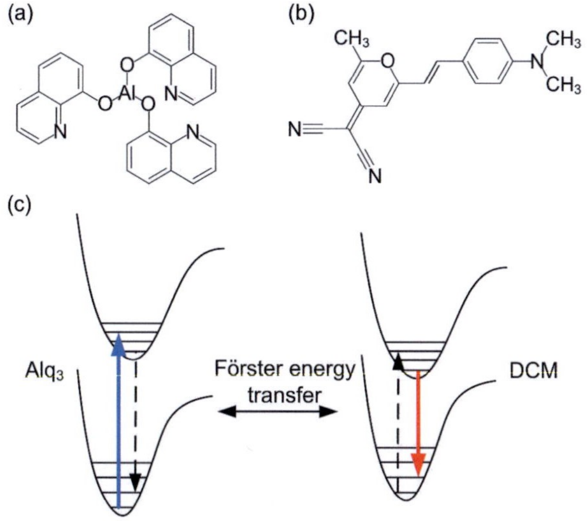

Figure 2.7: Chemical structures of (a) Alq_3 and (b) DCM. (c) Schematic illustration of host-guest Förster energy transfer between Alq_3 and DCM.

Alq_3 was used as a green emitter and gave rise to the transport of charge carriers in an *organic light emitting diode* (OLED) [33, 110, 111], while DCM is a commonly used laser dye having a high *photoluminescence quantum efficiency* (PLQE) close to 100 % and a large spectral tuning range due to its broad gain spectrum [67, 70]. When separated from DCM, Alq_3 has a PLQE as low as 35 %, and therefore direct lasing action from the undoped Alq_3 film has not been reported [112]. DCM exhibits no lasing effect in a neat thin film due to the concentration quenching [67]. However, enabling Förster energy transfer by doping Alq_3 with a very small amount of DCM significantly improves the overall efficiency and enables an emission shift from green to red (orange), as depicted in Figure 2.7(c). By varying the doping concentration of DCM from 0.5 wt.% to 10 wt.%, the photoluminescence wavelength can be tuned over a wide range from 605 nm to 650 nm [113, 114, 115].

The very large overlap in the green part of the spectrum between Alq_3 emission and DCM absorption leads to a very large Förster radius of 3.2 nm [112]. As the energy transfer time ($\sim 10^{-11}$–10^{-9} s) is much shorter than the radiative lifetime of Alq_3 ($\sim 1 \times 10^{-8}$ s) and DCM ($\sim 5 \times 10^{-9}$ s) [112], the composite small molecule can be considered as a four-level laser system. After doping, the absorption and the emission spectra of the whole small molecule system Alq_3:DCM exhibits a large Stokes shift of ~ 150 nm, making the system almost transparent to its own emission [112] (see Figure 2.8). The reduction of the optical loss originating from self-absorption leads to a very low laser threshold using Alq_3:DCM as active medium [21, 85, 112, 116, 117].

The refractive index of Alq_3:DCM varies from 1.7 to 1.78 in the emission spectral range of 600–650 nm and is about 1.72 at 633 nm [53, 118]. The stimulated emission cross section (σ_{SE}) of a Alq_3:DCM system is on the order of 10^{-17} cm^2 [53].

Both materials are commercially available (Alq_3 from Sensient, ST 1095S and DCM from Radiant Dyes) and the Alq_3:DCM thin film layer was deposited by thermal co-evaporation. The doping concentration was con-

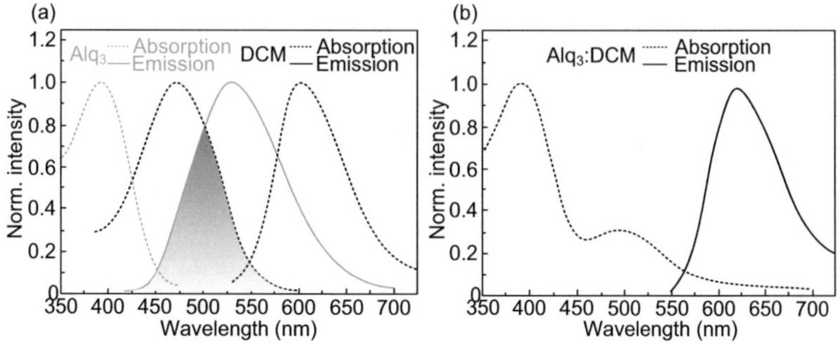

Figure 2.8: (a) Absorption and emission spectra of Alq_3 and DCM respectively. (b) The absorption and emission spectra of a composite Alq_3:DCM film with a DCM concentration of ~ 2 % by weight (adapted from [53]).

trolled within 2–3 % through adjusting the evaporation temperature of ~ 220 °C for Alq_3 and ~ 120 °C for DCM. With this doping concentration, the laser emission wavelength can be tuned from 600 nm to 725 nm by changing the resonator configuration or the film thickness [119, 120, 121].

The fabricated Alq_3:DCM laser device was characterized in a vacuum chamber or subsequently encapsulated in a nitrogen environment for ambient atmosphere characterization. The Alq_3:DCM laser lifetime was determined in vacuum to 9.8×10^7 laser pulses until the emission intensities dropped to 50 % of the initial value. There was no degradation in laser properties after a storage of up to 12 months in the dark in ambient atmosphere. The relatively high photostability of Alq_3:DCM over other organic semiconductor active media enables it to be an ideal candidates for spectroscopy and sensing applications, which require a long-time stability in laser output power and in laser emission wavelength.

The conjugated polymer $F8_{0.9}BT_{0.1}$

Fluorene-based polymers (polyfluorenes) appear to be a particularly attractive class of materials in applications of organic laser devices because of their high net gain. Green-yellow emitting poly[9,9-dioctylfluorenyl-2,7-

diyl)-co-1,4-benzo-{2,1-3}-thiadiazole)] (F8BT) is one of these promising fluorene copolymers. The optical characteristics of this copolymer are mainly determined by the different combinations of repeat units in its chemical structure (see Figure 2.9(a)). The F8-segment is the same as the backbone structure of poly[9,9-dioctylfluorene) (PFO), so this conjugated polymer is also called PFO-BT [98].

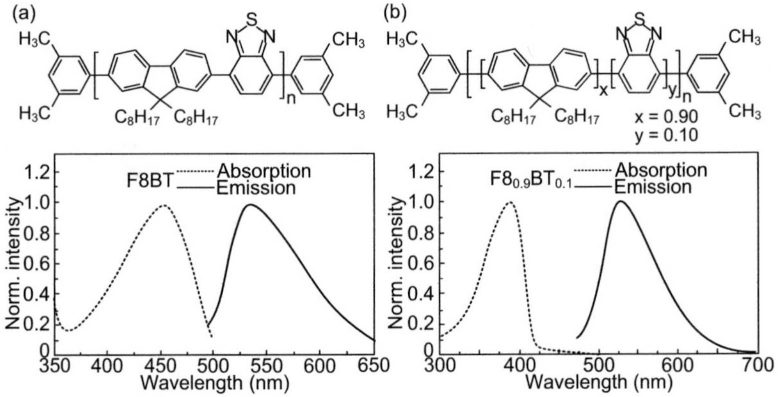

Figure 2.9: Chemical structure and emission/absorption spectra of (a) F8BT and (b) $F8_{0.9}BT_{0.1}$ in tetrahydrofuran (THF) (adapted from [122]).

F8BT shows a stimulated emission cross section of 7×10^{-16} cm^2 and a high net gain of 22 cm^{-1} in neat films [74]. Figure 2.9(a) shows the corresponding absorption and photoluminescence spectra, which indicate a strong absorption at $\sim$ 450 nm and a maximum emission at $\sim$ 535 nm. The optical gain of F8BT covers a spectrum range from 530 nm to 600 nm [123]. In this wavelength segment, the refractive index of F8BT neat film varies between 1.8 and 1.92 [124].

Due to a relatively low PLQE of $\sim$ 58 % in F8BT [74], an alternative polyfluorene derivative $F8_{0.9}BT_{0.1}$ with the same F8- and -BT units was used in this work with an improved PLQE of $\sim$ 80 % [23]. The chemical structure is shown in Figure 2.9(b). The high PLQE of $F8_{0.9}BT_{0.1}$ arises, because the Förster energy transfer occurs from the major F8-segments to the minor BT-segments [125]. The absorption and the refractive index of $F8_{0.9}BT_{0.1}$ is dominated by the F8-segments, i.e., the PFO-monomer. That is the reason why the absorption maximum shifts to a shorter wavelength at 360 nm (see Figure 2.9(b)) and the refractive index of $F8_{0.9}BT_{0.1}$ thin film is lower than the one of F8BT at the same emission wavelength [74, 98]. However, the emission range is still determined by the BT-segments and similar to that of F8BT.

Due to the large molecular weight, these two fluorene copolymers are not suitable for deposition by thermal evaporation, but can be solution-processed by spin-coating, horizontal dipping, ink-jet printing and other printing or coating techniques. F8BT and $F8_{0.9}BT_{0.1}$ used in this work were commercially available (American Dye Source: ADS133YE for F8BT and ADS233YE for $F8_{0.9}BT_{0.1}$), and highly soluble in toluene and tetrahydrofuran (THF). Toluene solutions of F8BT and $F8_{0.9}BT_{0.1}$ were made at a typical concentration of 20 mg ml^{-1}.

The composite conjugated polymer system $F8_{0.9}BT_{0.1}$:MEH-PPV

A promising conjugated polymer system for organic laser applications is the guest-host system comprising the host polyfluorene-derivative F8BT and the guest poly(p-phenylen-vinylen) (PPV)-derivative poly(2-methoxy-5-(2'-ethylhexyloxy)-1,4-phenylene vinylene) MEH-PPV [97] (see Figure 2.10(a)). The first demonstration of lasing from MEH-PPV in solution by Moses in 1992 has opened up a new research field of polymers as gain materials [126], but more significant is the investigation of lasing ability on neat MEH-PPV films [13, 127].

Casting a highly concentrated MEH-PPV solution into a thin film may cause an issue of polymer aggregation [128, 129], which leads to a lower stimulated emission cross section in film than in solution. The solvent, in which MEH-PPV is cast into films, plays an influential role in MEH-PPV photophysical properties. It was verified by Hide et al. [130] that a MEH-PPV film cast from THF solutions give a lower ASE threshold than that from chlorobenzene or *p*-xylene solutions. Since the rapid evaporation of the relatively low-boiling THF solvent tends to favor retention of the tightly coiled chains of MEH-PPV [13, 128], there is no adequate time for chain-chain or substrate-chain interactions to cause polymer aggregation before the solvent evaporates.

The strongest absorption of MEH-PPV (ADS100RE, American Dye Source) is at ~ 490 nm in film and the maximum photoluminescence is at ~ 585 nm (see Figure 2.10(b)). The $F8_{0.9}BT_{0.1}$ fluoresces at the wavelength near the absorption maximum of MEH-PPV. With a much lower concentration of MEH-PPV than in the sole recipe, the guest-host blend improves the film morphology. Again, Förster energy transfer, where the excitation is transferred from $F8_{0.9}BT_{0.1}$ (host) to MEH-PPV (guest), leads to very low laser thresholds [97, 108]. The lasing wavelengths in the $F8_{0.9}BT_{0.1}$:MEH-PPV system are determined by the MEH-PPV photoluminescence spectrum. With a broad gain spectrum from 592 nm to 690 nm [98], this polymer composite is a good candidate for spectrally tunable organic lasers [22, 131].

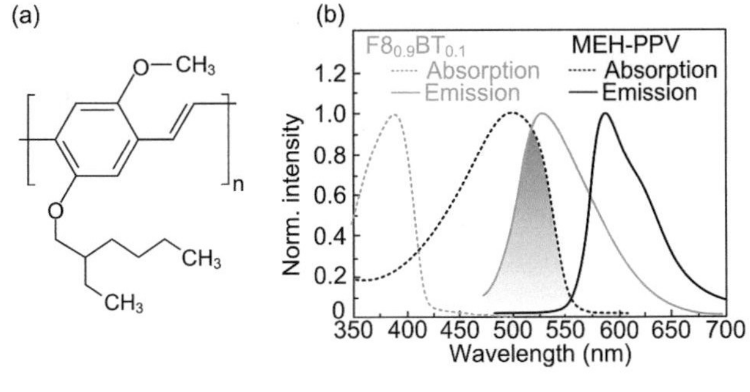

Figure 2.10: (a) Chemical structure and (b) emission/absorption spectra of MEH-PPV in THF in comparison to the spectra of $F8_{0.9}BT_{0.1}$ (adapted from [122]).

Similar to $F8_{0.9}BT_{0.1}$, the polymer blend $F8_{0.9}BT_{0.1}$:MEH-PPV can be deposited by solution processes. The blend concentration plays an important role in the efficient energy transfer. As reported in Ref. [97], a minimum threshold occurs at a proportion of 85:15 wt.%. A concentration of 20 mg ml^{-1} in toluene is used in this work.

2.2.5 Pump sources for organic semiconductor lasers

Challenges on electrically pumped organic lasers

An electrically pumped organic laser, i.e., an organic laser diode, has attracted a lot of interest since the demonstration of optically pumped organic lasers. Numerous efforts have been made to potentially solve the existing problems as well as few demonstrations have been presented to claim lasing under electrical pumping [132, 133, 134, 135, 136]. However, until now no convincing organic laser diode has been reported [66]. Some main issues remain as the obstacles in front of this goal.

First, different from singlet excited states in optical excitation, the charge carriers involved in electrical excitation, also referred to as polarons, have broad absorption spectra overlapping with the emission spectra and may quench the singlet excitons. Another issue is the triplet formation. Under optical pumping, the triplet states originate only through intersystem crossing and the triplet accumulation is not relevant by applying short-pulse excitation. However under electrical pumping, triplet excitons are created with a probability ratio of 75 % [137]. It will facilitate the unfavorable optical loss process, e.g., triplet-triplet absorption.

There are also extra losses, which are not present in optical pumping. Due to the losses associated with metallic electrodes, the waveguide confinement is strongly weakened and the actual laser threshold is greatly increased [138, 139]. Meanwhile, a high current density is correspondingly required to reach the laser threshold. Owing to the low charge carrier mobility, organic semiconductors can not continuously sustain the current densities as high as several kA cm^{-2}, since they may overheat and be destroyed [68].

The electrode contact issue can be relieved with careful configuration designs by reducing the overlap between waveguide mode and electrodes, which have been accomplished in several ways [140, 141, 142]. But the fundamental key to all issues mentioned above is the improvement of organic materials in aspects of charge carrier mobility, triplet lifetime, absorption cross section of excited states and so on.

Optical pump sources for organic lasers

Until now, organic solid-state lasers are pumped optically with short pulsed light sources. In addition to the demand of high stability, the emission spectra of optical pump sources should match the organic gain media for a high spectral absorption. For pulsed optical pumping, the pulse duration is usually longer than the laser emission lifetime [98], but shorter than the excited state (e.g., triplet) lifetime and intersystem crossing time constant [68, 98]. An optimum optical excitation pulse has a duration as short as possible to allow an instantaneous population inversion [69].

Diode pumped solid-state (DPSS) lasers are qualified candidates as optical pump sources for organic semiconductor lasers thanks to their high conversion efficiency [64, 65]. Usually, the doped crystal neodymium-doped yttrium aluminum garnet (Nd:YAG) or a neodymium-doped yttrium orthovanadate (Nd:YVO_4) is pumped by a powerful infrared GaAlAs laser diode with a wavelength of 808 nm. The doped crystal produces 1064 nm lasing from the main spectral transition of the neodymium ion. This light is then frequency-multiplied using a nonlinear optical process in a nonlinear crystal, e.g., potassium titanyl phosphate (KTP) producing 532 nm or 355 nm laser emission. DPSS lasers operate in both pulsed and continuous modes. With the help of the Q-switching mode, a pulse duration < 1 ns can be achieved [64]. Additionally, these lasers can reach a very high power while maintaining a relatively good beam quality.

To reduce apparatus costs and size of the system, there are plausible ways to achieve optical pumping using smaller and cheaper light sources. Among them, a violet or blue emitting pulsed (In)GaN laser diode has been verified as efficient pump source for organic semiconductor lasers based on both conjugated polymers [19, 20, 22] and small molecules [21]. Aiming for even lower costs, light emitting diodes (LEDs) have also been successfully applied for optical pumping of organic lasers [23, 24, 25, 26], allowing the advent of a hybrid organic-inorganic laser diode. Since LEDs are incoherent light sources and the pulse duration cannot be defined as short as laser diodes, the power density is just enough to slightly overcome the lasing threshold of organic semiconductor lasers [23]. The application fields mainly focus on those that have no requirement for the laser output power, like sensing, but not on those that care about the output power capability.

2.3 Organic semiconductor distributed feedback lasers

2.3.1 Laser resonators

Since the first introduction of an optically pumped organic microcavity laser in 1996 [12], optically pumped organic semiconductor lasers have been demonstrated in a wide variety of different resonant structures. Figure 2.11 features most of the geometries that have been explored in the past years using optical excitation.

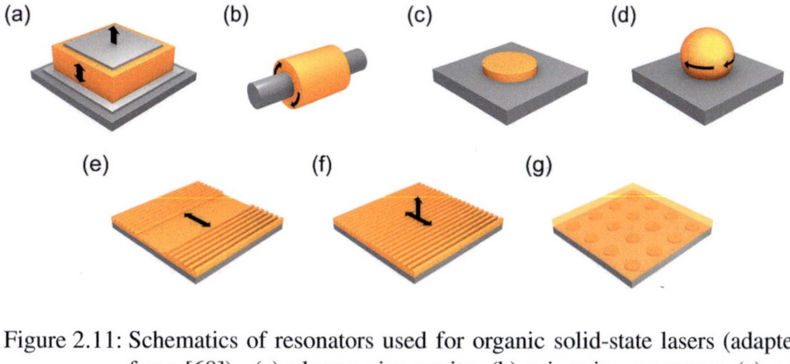

Figure 2.11: Schematics of resonators used for organic solid-state lasers (adapted from [68]): (a) planar microcavity, (b) microring resonator, (c) microdisk resonator, (d) microsphere resonator (e) distributed Bragg mirror on both sides of a waveguide resonator, (f) distributed feedback resonator and (g) two-dimensional photonic crystal.

In contrast to the cavities based on bulk gain media for solid-state dye lasers, e.g., the conventional Fabry-Pérot laser, resonators for organic semiconductor lasers are usually designed with a thin-film concept. Organic semiconductors can be easily deposited as thin films on various substrates or cavity surfaces. The thicknesses and the shapes of the thin layers can be manipulated in a relatively simple way through thermal evaporation and solution processing techniques.

The resonators for organic semiconductor lasers comprise two major categories: microcavities and waveguide resonators.

The most simple form of a microcavity is the *planar microcavity*, which is inspired by the stacked multilayers in OLEDs and the inorganic *vertical-cavity surface-emitting laser* (VCSEL) [12]. It consists of an organic layer between two mirrors forming a Fabry-Pérot microcavity, as shown in Figure 2.11(a). One mirror has a high reflectivity at the laser emission spectra range and the other has a partial transmission which functions as an output coupler. The planar microcavity was firstly used for a conjugated polymer PPV laser [12] and later also successfully applied to many other materials [37, 143, 144, 145, 146]. This configuration is also known as organic vertical-cavity surface-emitting laser (OVCSEL) geometry [109, 145, 146, 147]. Due to the low roundtrip gain associated with the short gain length, the laser threshold of these devices is relatively high [69]. This issue can be resolved by using the following microcavities or waveguide resonators.

The simple processing of organic semiconductors enables them to be fabricated in novel ways that are not possible for inorganic semiconductors. Good examples are *microring* [148, 149, 150, 151], *microdisk* [40, 152, 153] and *microsphere* [154] lasers (as shown in Figure 2.11(b)–(d)), which are based on the principle of whispering-gallery-modes originating from the total internal reflection at the interface between the organic active material and the surrounding medium [155, 156]. Thanks to low losses in the cavities, very low laser thresholds can be achieved using such microcavities. By changing the resonator shape, the laser emission from these ring-like microcavities can be tailored as regular comb-like spectra [157] or as chaotic output [158], depending on the total reflection and radiation directions.

An alternative way of realizing an organic laser is using waveguide resonators. In this geometry, the resonator axis is parallel to the film plane, which allows a long gain length during a roundtrip. A straightforward arrangement is a Fabry-Pérot waveguide. The light is waveguided in the high refractive index organic film via total internal reflection interface. Due to the difficulty in forming good quality edges with polymer films, an alternative mirror configuration using *distributed Bragg reflector* (DBR) gratings is used to replace the end facets [19, 107, 159], as depicted in Figure 2.11(e).

The efficiency of waveguide lasers can be further improved by diffractive structures, which do not rely on mirrors, but on periodic, wavelength-scale corrugations that diffract or Bragg-scatter the light in the spectral regions of optical gain. There have been many different diffractive structures explored for organic semiconductor lasers, including *distributed feedback* (DFB) structure [74, 160, 161, 162, 163] (see Figure 2.11(f)) and two- or three-dimensional photonic crystal structures [164, 165, 166] (see Figure 2.11(g)). DFB lasers have been proven to be particularly successful and will be discussed in more detail in the following sections.

2.3.2 Distributed feedback resonators

Slab Waveguides

Figure 2.12 shows a dielectric three-layer slab waveguide consisting of a thin film, sandwiched between a substrate and a cladding. The propagation of light in an isotropic homogeneous dielectric medium can be described by the Helmholtz equation. A mode of the dielectric waveguide is a solution of the wave equation:

$$\nabla^2\mathbf{E}(\mathbf{r}) + k_0^2 n^2(\mathbf{r})\mathbf{E}(\mathbf{r}) = 0 \qquad (2.6)$$

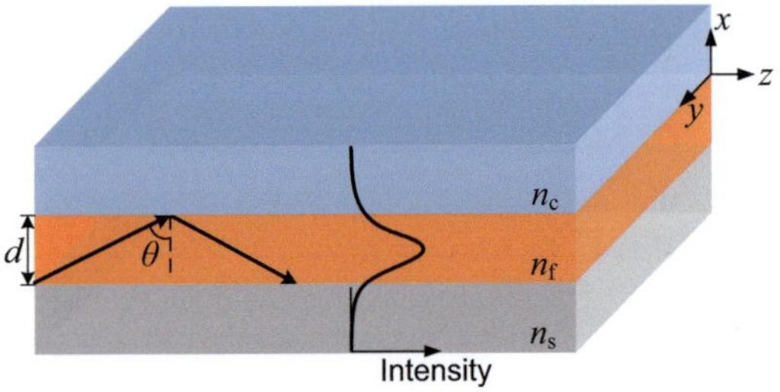

Figure 2.12: Schematic of a typical dielectric three-layer waveguide.

where $k_0 = 2\pi/\lambda$ is the magnitude of the wave vector **k** (also known as wavenumber) and n is the refractive index. The solution takes the form of

$$\mathbf{E}(\mathbf{r},t) = \mathbf{E}(x,y)e^{i(\omega t - \beta z)} \tag{2.7}$$

so that Eq. 2.6 becomes

$$(\frac{\partial^2}{\partial x^2} + \frac{\partial^2}{\partial y^2})\mathbf{E}(x,y) + [k_0^2 n^2(x,y) - \beta^2]\mathbf{E}(x,y) = 0 \tag{2.8}$$

The confined modes in a film are only possible when $k_0 n_c, k_0 n_s < \beta < k_0 n_f$, i.e., $n_f > n_c, n_s$, which means the inner layer possesses the highest index of refraction. $n_{eff} = \beta \frac{c}{\omega} = \beta \frac{\lambda}{2\pi}$ is denoted as effective refractive index, which satisfies n_c, $n_s < n_{eff} < n_f$. $E(x)$ can be solved according to the boundary conditions while decaying exponentially in cladding and substrate [167].

There are two cases for electromagnetic waves in a waveguide: TE (transverse electric) waves with the electric field vector **E** perpendicular to the propagation direction and TM (transverse magnetic) waves with the magnetic field vector **H** perpendicular to the propagation direction. In the context of semiconductor lasers, mostly TE modes are considered, as TE modes are usually stronger localized in the waveguide layer and thus experience larger gain [49, 167]. Thus, the discussion focuses on TE modes. The treatment of TM modes, however, is analogous.

For a symmetric slab waveguide ($n_c = n_s$), a fundamental mode can exist at any frequency, while all the other modes with higher orders can exist only above a "cutoff" frequency. And if the slab waveguide is asymmetric ($n_c \neq n_s$), the fundamental mode also possess a cutoff frequency [167].

The TE mode solutions for an asymmetrical slab waveguide with a film thickness of d can be taken as [168]

$$E_y(x) = A \begin{cases} \exp(-qx) & 0 \leq x < \infty \\ \cos hx - \frac{q}{h}\sin hx & -d \leq x \leq 0 \\ (\cos hx + \frac{q}{h}\sin hx)\exp[p(x+d)] & -\infty < x \leq -d \end{cases} \tag{2.9}$$

with

$$h = \sqrt{n_f^2 k_0^2 - \beta^2} \tag{2.10}$$

$$q = \sqrt{\beta^2 - n_c^2 k_0^2} \tag{2.11}$$

and

$$h = \sqrt{\beta^2 - n_s^2 k_0^2} \tag{2.12}$$

The coefficients are determined by making $E_y(x)$ and $\partial E_y(x)/\partial x$ continuous at both interfaces $x = 0$ as well as $x = -d$. So a simplified equation is obtained as

$$\tan ht = \frac{p+q}{h(1 - pq/h^2)} \tag{2.13}$$

The eigenvalues of n_{eff} can be obtained for the confined TE modes. Figure 2.13 shows an example of the effective refractive index as a function of film thicknesses for a wavelength of 630 nm in a typical organic film ($n_f = 1.7$) deposited on a planar glass substrate ($n_s = 1.5$).

One-dimensional DFB lasers

Figure 2.14 shows the typical structure of organic semiconductor DFB lasers (DFB-OSLs) with diffractive feedback along the plane axis of the waveguide. The laser consists of a thin organic semiconductor layer de-

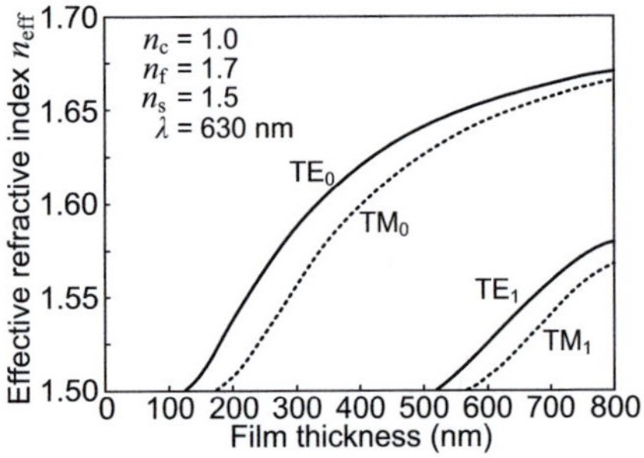

Figure 2.13: Effective refractive index as function of film thickness for a wavelength of 630 nm in a typical organic film deposited on a planar glass substrate.

posited on top of a corrugated substrate. Light propagates as a waveguide mode mainly in the organic active layer, which has a higher refractive index than the substrate.

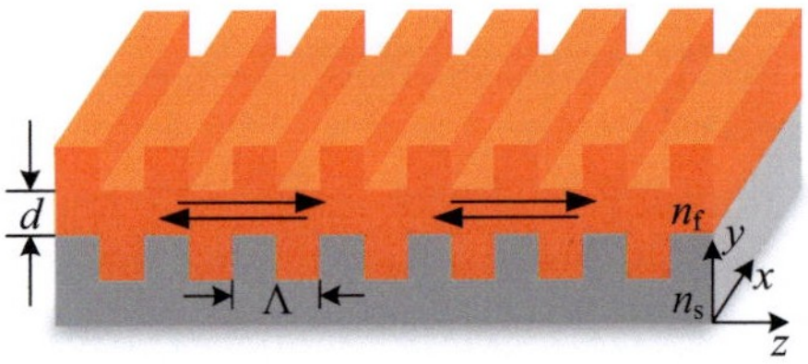

Figure 2.14: Schematics of an organic DFB laser with corrugations of period Λ and an active layer with thickness of d.

For a given period of the corrugation, there is a particular set of wavelengths that will be diffracted from a propagating mode of the waveguide into the counter-propagating waveguide mode. This situation will arise when the Bragg condition is satisfied:

$$m\lambda_{\text{Bragg}} = 2n_{\text{eff}}\Lambda \tag{2.14}$$

where λ_{Bragg} is the Bragg wavelength, Λ is the periodicity of the corrugation structures, and m is an integer that represents the order of diffraction. n_{eff} is the so-called effective refractive index of the waveguide as discussed above. It may be calculated through a solution of the Helmholtz wave equation for a planar multilayer structure as described above. The DFB laser emission wavelength can be estimated as $\lambda_{\text{Laser}} \approx \lambda_{\text{Bragg}}$, which is only valid when the Bragg wavelength lies within the gain spectral range of the employed active medium.

As shown in Figure 2.15, for first-order corrugation structures $p = 1$, where p is the grating order, the laser radiation is emitted from the edges of the structure and the laser wavelength is $\lambda_{\text{Laser}} = 2n_{\text{eff}}\Lambda$. This scheme provides low-threshold laser operation and facilitates an in-plane light coupling into a waveguide [169, 170, 171], i.e., an integrated laser in a lab-on-a-chip system is possible [18, 30, 120]. However, it always suffers from poor laser beam quality due to the rough edge of the amorphous organic layers. Moreover, as the thickness of the employed organic semiconductor films is usually in the range of few hundred nanometers, the resulting beam divergence is very large. Furthermore, first-order gratings require even smaller corrugation features, e.g., for a blue-emitting laser at a wavelength of 450 nm, a small grating periodicity $\sim$ 120 nm and a feature size down to $\sim$ 60 nm is required, which makes it very difficult to reach a high fabrication accuracy with conventional photolithography technique [49].

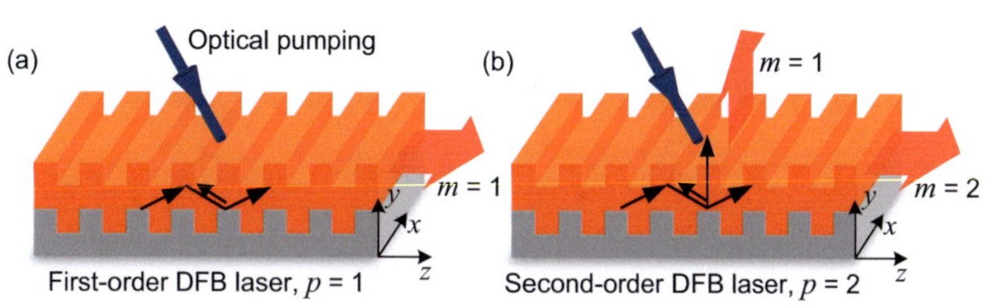

Figure 2.15: Scheme of a DFB laser with feedback in (a) first and (b) second grating order.

For the case of second-order corrugation structures $p = 2$, the laser emission wavelength λ_{Laser} equals $n_{\text{eff}}\Lambda$. In addition to edge-emission, the laser radiation is also emitted perpendicular to the plane of the waveguide [172]. Such structures can therefore provide a surface-emitting output coupling of the laser light via first-order diffraction while providing in-plane feedback via second-order diffraction, since if the Bragg condition is fulfilled with $p \geq 2$ then all the lower diffraction orders m will be scattered out of the waveguide [172].

To combine the advantages of the first-order and the second-order DFB lasers, a mixed-order DFB laser has been proposed [173, 174]. It comprises first-order regions providing strong feedback and a second-order region for perpendicular outcoupling. An optimum trade-off between strong outcoupling and low threshold is realized by optimizing the grating widths and organic film parameters.

For a more detailed investigation of DFB lasers, the wave equation 2.6 should be solved in the presence of the periodic modulation. The coupled wave theory can be used to elucidate the DFB periodic waveguide, which was established in 1972 by Kogelnik and Shank [175] and supplemented in 1977 by Streifer et al. [176].

It is assumed that a spatial modulation of the refractive index $n(z)$ and of the gain constant $g(z)$ are of the form [175]

$$n(z) = n + \Delta n \cos 2\beta_0 z \tag{2.15}$$

$$g(z) = g + \Delta g \cos 2\beta_0 z \tag{2.16}$$

where n and g are the average values of the medium parameters and Δn and Δg are the amplitudes of the spatial modulation. With these conditions the wavenumber k can be rewritten in the form

$$k^2 = \beta^2 + 2ig\beta + 4\kappa\beta \cos 2\beta_0 z \tag{2.17}$$

where $\beta = n\omega/c$, $\beta_0 = n\omega_0/c$. The coupling constant κ is defined by

$$\kappa = \pi\Delta n/\lambda_0 + \frac{1}{2}i\Delta g \tag{2.18}$$

where λ_0 is the wavelength corresponding to the Bragg frequency ω_0.

The two significant waves in the DFB structure are two counter-running waves L (left traveling wave) and R (right traveling wave). These waves are described by the complex amplitudes $L(z)$ and $R(z)$ and yield the electric field as the sum

$$E(z) = L(z)\exp(i\beta_0 z) + R(z)\exp(-i\beta_0 z) \tag{2.19}$$

Substituting Eq. 2.19 into Eq. 2.6, a pair of coupled-wave equations can be obtained

$$\frac{\partial L}{\partial z} + (g - i\delta)L = i\kappa R \tag{2.20}$$

$$-\frac{\partial R}{\partial z} + (g - i\delta)L = i\kappa L \tag{2.21}$$

The parameter δ is the detuning from the Bragg frequency, defined by

$$\delta \equiv \frac{(\beta^2 - \beta_0^2)}{2\beta} \approx \beta - \beta_0 = n(\omega - \omega_0)/c \tag{2.22}$$

With the boundary conditions for the wave amplitudes in a structure of length l: $R(-\frac{1}{2}l) = L(\frac{1}{2}l) = 0$, the corresponding solutions are

$$L = l_1 e^{\gamma z} + l_2 e^{-\gamma z} \tag{2.23}$$

$$R = r_1 e^{\gamma z} + r_2 e^{-\gamma z} \tag{2.24}$$

with the complex propagation constant γ obeying the dispersion relation

$$\gamma^2 = \kappa^2 + (g - i\delta)^2 \tag{2.25}$$

Eq. 2.25 defines the oscillation condition, i.e., a gain threshold for a standing wave with a wavenumber of β. Based on this result, the mode distribution of the two counter-running waves can be solved.

Because of the assumed symmetry of the device, the coefficients l_i and r_i are related by

$$l_1 = \pm r_2 \tag{2.26}$$

$$l_2 = \pm r_1 \tag{2.27}$$

Using these results, the longitudinal field distribution of the modes of a DFB structure are described as follows

$$L = \pm \sinh \gamma (z - \frac{1}{2} l) \tag{2.28}$$

$$R = \sinh \gamma (z + \frac{1}{2} l) \tag{2.29}$$

In the case of index modulation ($\Delta g = 0$), there is no oscillation taking place exactly at the Bragg wavelength λ_{Bragg} given by 2.14 [167]. Two laser modes can be found at wavelengths which are symmetrically separated from the Bragg wavelength. It forms a so-called stopband or band gap between two laser wavelengths with a distance of

$$\Delta\lambda_{\mathrm{stop}} \approx \frac{\kappa \lambda_{\mathrm{Bragg}}^2}{\pi n_{\mathrm{eff}}} \tag{2.30}$$

The threshold gain value is obtained from

$$\frac{e^{2 g_{\mathrm{th},m} L}}{g_{\mathrm{th},m}^2 + \delta_m^2} = \frac{4}{\kappa^2} \qquad m = 0, \pm 1, \pm 2, \ldots \tag{2.31}$$

It indicates an increase in threshold with increasing mode number m. Two lowest threshold modes (with $m = 0$ and $m = -1$) are situated symmetrically on either side of the Bragg wavelength λ_{Bragg} just outside the band gap. In practice, this situation is not desirable, because it results in wavelength instability and spectral broadening. To improve this property, an extra section with period $\Lambda/2$ is widely employed at the center of the laser cavity [167, 177]. This added propagation delay will "force" the DFB laser oscillating preferentially at a single midgap wavelength.

In the case of gain modulation ($\Delta n = 0$), the laser oscillation takes place at the exact Bragg wavelength, which means that there is no frequency stop band for gain coupling. When two modulations exist at the same time, the presence of gain coupling will lift the degeneracy of the modes and thus favor monomode operation [178].

Laser beam quality and far-field laser emission profiles are also important properties for free-space applications. Thanks to the strict mode selectivity, the DFB laser is in principle a longitudinal single-mode laser and they can emit a clean monochromatic light at the Bragg wavelength. However, due to the absence of lateral confinement, the one-dimensional (1D) DFB output beam is not in the circular shape of an excitation spot, but often a divergent fan. This effect can be directly observed in the emission pattern of second-order DFB lasers, as shown in Figure 2.16. Owing to an additional wavevector component parallel to the corrugation orientation $\mathbf{k}_x$, the

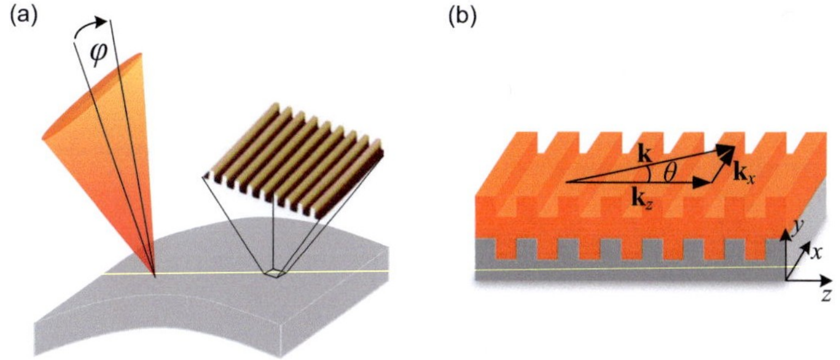

Figure 2.16: (a) Schematic illustration of the fan-like laser emission beam observed in a second-order 1D-DFB laser above threshold. (b) Scheme of the wavevector components of the laser modes in a 1D-DFB laser (adapted from [116]).

distributed feedback is not only provided for wavevectors perpendicular to the grating $\mathbf{k}_z$, but also for the waves propagating at an angle θ relative to $\mathbf{k}_z$. It leads to a fraction of the far-field laser beam emitting at an angle φ relative to the surface normal and an angular dependent laser wavelength [116]:

$$\lambda(\varphi) = n_{\text{eff}}\Lambda\sqrt{1-\left(\frac{\sin\varphi}{n_{\text{eff}}}\right)^2} \tag{2.32}$$

The lateral modes increase the spectral width of the laser line (FWHM) [116]. The higher lateral modes exhibit higher laser thresholds and they contribute significantly to the overall laser emission at high excitation densities [53, 179]. According to Eq. 2.32, the laser wavelength decreases with increasing output angle. Thus, the spectral broadening due to more excited lateral modes is in shorter-wavelength direction. This effect will be described in detail in Section 5.2.3.

Two-dimensional DFB lasers

In addition to the basic one-dimensional DFB laser, a two-dimensional (2D) feedback in the plane of an organic semiconductor film applied by complicated diffractive resonators can be introduced. These structures commonly form a 2D photonic crystal with either square [169], hexagonal [180], honeycomb lattices [164] or novel concentric circular shape [181, 182].

The quantitative theoretical investigations of 2D-DFB lasers have been presented for rectangular gratings [183, 184, 185, 186, 187]. Based on the theory of Sakai et al. [187], the 2D-grating structure can be considered as a square lattice of circular holes in the $x-y$ plane with period of $\Lambda_x = \Lambda_y = \Lambda$. The reciprocal lattice is also rectangular with period of $G_{0,x} = G_{0,y} = 2\pi/\Lambda$, as shown in Figure 2.17.

In such resonators, feedback may be applied in several directions in the plane of the film. In the case of a square lattice with reciprocal lattice vectors $\mathbf{G}$, there are eight waves propagating in the lattice structure and interfering with each other due to the diffraction by the circular holes. Among

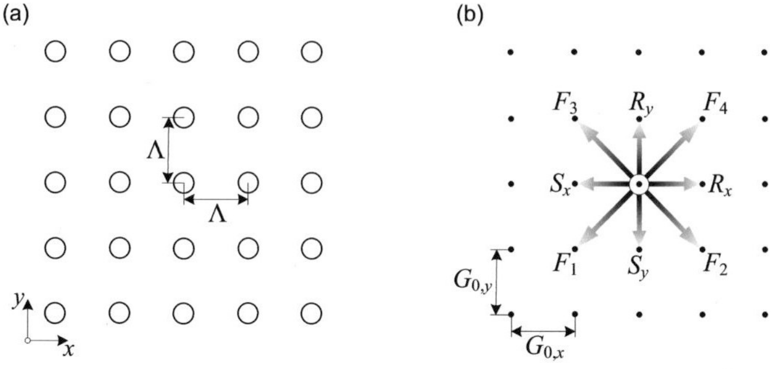

Figure 2.17: Schematics of rectangular lattice. (a) Real space rectangular lattice. (b) The corresponding reciprocal lattice and wave propagation (adapted from [187]).

them, R_x, S_x, R_y and S_y are the complex amplitudes of the waves propagating along the x and y directions, whereas F_1-F_4 are the complex amplitudes of the waves propagating along the 45° direction between x and y orientations, as shown in Figure 2.17(b). The corresponding coupling constants are as follows: κ_0 describing the diffraction in the direction vertical to the grating plane, $\kappa_1 = \kappa_{\mathbf{G}_1}$, where $|\mathbf{G}_1| = \beta_0$ and $\beta_0 = 2\pi/\Lambda$, describing the intensity of the coupling of two plane waves propagating at 45° to each other and $\kappa_2 = \kappa_{\mathbf{G}_2}$, where $|\mathbf{G}_2| = 2\beta_0$, describing the intensity of the coupling of counter-propagating waves, which corresponds to the backward scattering in 1D-DFB lasers. The coupled-wave equations for two-dimensional TE modes can be written as [187]

$$-\frac{\partial}{\partial x}R_x + (g - \kappa_0 - i\delta)R_x = i\frac{4\kappa_1^2}{\beta_0}R_x + (i\kappa_2 - \kappa_0)L_x + i\frac{2\kappa_1^2}{\beta_0}(L_y + R_y) \tag{2.33}$$

$$\frac{\partial}{\partial x}L_x + (g - \kappa_0 - i\delta)L_x = i\frac{4\kappa_1^2}{\beta_0}L_x + (i\kappa_2 - \kappa_0)R_x + i\frac{2\kappa_1^2}{\beta_0}(L_y + R_y) \tag{2.34}$$

$$-\frac{\partial}{\partial y}R_y + (g - \kappa_0 - i\delta)R_y = i\frac{4\kappa_1^2}{\beta_0}R_y + (i\kappa_2 - \kappa_0)L_y + i\frac{2\kappa_1^2}{\beta_0}(L_x + R_x) \tag{2.35}$$

$$\frac{\partial}{\partial y}L_y + (g - \kappa_0 - i\delta)L_y = i\frac{4\kappa_1^2}{\beta_0}L_y + (i\kappa_2 - \kappa_0)R_y + i\frac{2\kappa_1^2}{\beta_0}(L_x + R_x) \tag{2.36}$$

Until now, it has been demonstrated that 2D-DFB lasers may exhibit improved output beam quality, lower threshold and higher slope efficiencies than their 1D-counterparts [166, 169, 186, 188]. For a surface-emitting 1D-DFB laser, the output beam is emitted as a divergent stripe, parallel to the orientation of the grating grooves (see Figure 2.18(a)). In the 2D-

Figure 2.18: The emission patterns of Alq_3:DCM DFB lasers pumped by a Nd:YVO_4 Laser (355 nm): (a) 1D-DFB laser. (b) 2D-square-array DFB laser pumped with pulse energy 2 times higher than the laser threshold. (c) 2D-square-array DFB laser pumped with pulse energy 5 times higher than the laser threshold.

structure, at modest excitation densities, the output beam has a central spot with divergence of typically only a few milliradians in any direction and is close to being diffraction limited (see Figure 2.18(b)) [53, 68, 69]. At higher pump power, the emission forms oftentimes a cross shape along the symmetric axes of the grating, as illustrated in Figure 2.18(c).

2.3.3 Applications

In parallel to the fundamental investigations and advancements of organic semiconductor DFB lasers (DFB-OSLs), numerous real-life applications have been proposed for exploiting the attractive properties of these novel laser emitters. Taking advantages of simple fabrication and broad spectral emission range, OSLs can be utilized in specific novel application fields or may replace coherent emitting sources in traditional laser applications. It has been found that conjugated polymer lasers could be applied in biosensing for specific antibody or gas phase molecules [189, 190] and in chemical sensing of explosive vapor such as 2,4,6-trinitrotoluene (TNT) and 2,4-dinitrotoluene (DNT) [191, 192, 193]. Compared to conventional fluorescence detection schemes, the modification on laser emission wavelength, laser slope efficiency or laser threshold reflecting the adsorbed analyte molecules has a much higher sensing sensitivity [191]. In the field of optical telecommunication, organic semiconductor gain media can act as optical amplifiers [194, 195] and resonant switching elements [196, 197]. Most promising of all, thanks to the broad gain spectrum and moderate pulse energy, organic semiconductor lasers are well suited as spectroscopic light sources with potential for fluorescence excitation, absorption/transmission spectroscopy, Raman spectroscopy, etc. Below, the applications in spectroscopy are briefly reviewed.

Spectrally tunable organic DFB lasers

The major advantage of organic lasers over conventional inorganic solid-state lasers is the wide continuous wavelength tunability, which plays an important role in the application of spectroscopy. It has been found that a fine tuning of laser wavelength could be achieved by changing the temperature of the organic active layer [198, 199, 200]. Moreover, according to Eq. 2.14, the laser emission wavelength tunability can be achieved by varying the period of DFB-corrugation or the effective refractive index of DFB-waveguide.

First, DFB lasers are simply tunable upon changing the grating period, which can be achieved by discrete steps [120, 201] or by using a specific fabrication process to create continuously varying grating period, e.g., Wang et al. used multiple e-beam exposure in fabricating chirped gratings with 40 nm period variation, corresponding to an organic DFB laser with a wavelength tunability of 38 nm [119]. Another novel idea for continuous grating variation is using a flexible grating substrate, e.g., the common elastomer polydimethylsiloxanc (PDMS) or an electroactive elastomer. Triggered by a mechanical deformation or a voltage variation, the grating period on the flexible substrate can be continuously shifted. This approach has been used in organic lasers with different active media [202, 203, 204, 205, 206]. A maximum tunability of 47 nm was achieved [206].

The DFB laser emission wavelength can also be tuned by changing the effective refractive index of the entire laser waveguide n_{eff} [207], which is dependent on the refractive index of active layer itself n_{f}. One approach to directly change the layer refractive index is introducing an optofluidic laser with organic laser dyes. Through successive mixture of solvents with different refractive indices, the optofluidic slab refractive index changes as well [208, 209]. Additionally, the film thickness of the active layer also determines n_{f}, which has been verified by measuring the laser emission wavelength on samples with different layer thicknesses [85]. It is more elegant to realize a continuously tunable laser on one device via creating a wedge-shaped active layer, as illustrated in Figure 2.19. For small molecules, the

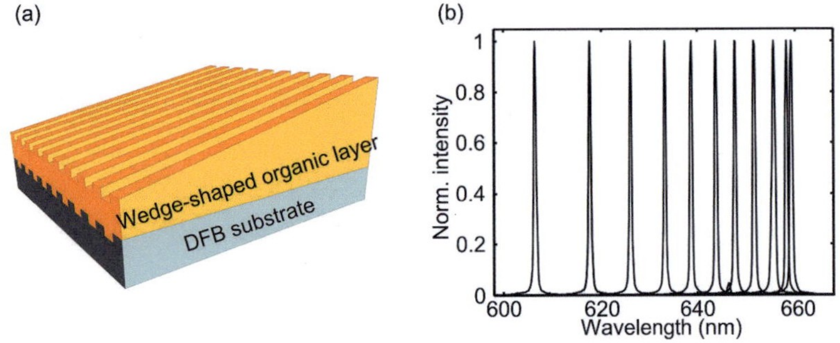

Figure 2.19: (a) Schematic illustration of a DFB laser with a wedge-shaped organic active layer. (b) Corresponding laser spectra measured at different positions on a DFB laser with a wedge-shaped active layer of Alq_3:DCM [121].

film thickness gradient can be achieved by evaporating the active medium through a rotating shadow mask [121]. This method has also been used in fabricating tunable organic microcavity lasers [210] and graded-reflectance mirrors [211]. Using shadow masks with different shapes, a maximum tuning range up to 55 nm was obtained using Alq_3:DCM as the active medium [212]. For conjugated polymers, as introduced in Section 3.2.2, a pre-metered coating process named horizontal dipping can be applied for depositing a wedge-shaped organic active layer. A position-dependent film thickness on the sample is given by accelerated/decelerated transport of the coating barrier along the sample. Using this method, a reproducible wavelength tunability as wide as 19 nm was achieved using $F8_{0.9}BT_{0.1}$:MEH-PPV as the composite active medium [22].

The effective refractive index n_{eff} of the propagating slab waveguide mode also depends on the refractive indices of ambient layers embracing the active layer. Some chemical and biosensing schemes with organic lasers are based on this principle: The variation in laser emission wavelength is due to the changing n_{eff} during adsorption of the analyte molecules [190]. By adding a high refractive index intermediate layer with a wedge-shaped thickness, the organic laser wavelength can also be continuously tuned over 20 nm [213]. The refractive index contrast can furthermore be achieved by introducing a novel DFB laser design incorporating a voltage-controlled liquid crystal (LC) cell adjacent to the active layer [214, 215]. By increasing the electrical field between electrodes aligned to the LC, a shift of lasing wavelength can be observed. The wavelength tuning is a dynamic procedure, which is related to the thickness of LC cell, the distance between two electrodes and the instantaneous voltage signal regulation [98, 215].

Spectroscopic applications

Since the demonstration of Schneider et al. in 2005 of a UV tunable organic DBR laser as excitation source for biomarker experiments [27], organic solid-state lasers have attracted a considerable attention as potential light sources for spectroscopic applications. Inheriting the advantages from their original organic dye lasers, organic solid-state lasers offer widely tunable laser radiation over the UV and the visible spectral range and may contribute this unique advantage in spectroscopy.

Up to now, organic semiconductor lasers have been successfully applied in fluorescence excitation [27], absorption spectroscopy [28], and transmission spectroscopy [29, 212].

2.4 Raman spectroscopy

Since the first reports in 1928 [3, 4], Raman spectroscopy has played an increasingly important role in chemistry, biophysics and life science. The advent of the laser as an intense and monochromatic source of excitation light improved the Raman signals dramatically.

Raman spectroscopy is an ideal label-free optical detection technique which can provide rich information on the molecular structure and composition of analytes. In this section, the fundamentals of Raman scattering and the excitation sources for laser Raman spectroscopy will be introduced. Furthermore, a detailed description of surface-enhanced Raman spectroscopy as well as its applications is given.

2.4.1 Raman scattering

The Raman effect was theoretically predicted in 1923 by A. Smekal [216], and experimentally demonstrated in 1928 by C. V. Raman and K. S. Krishnan in liquids [3] and, independently, by L. I. Mandelstam and G. Landsberg in crystals [4]. In terms of the particle aspect of light, the Raman effect can be described as an inelastic scattering process between a photon and a molecule. When photons are scattered from a molecule, elastic scattering dominates with no gain/loss of energy, and therefore no frequency change. This is known as Rayleigh scattering. Raman scattering, however, is an inelastic scattering of the incident photons whereby energy is transferred to or received from the sample due to changes in the vibrational or rotational modes of sample molecules, causing a change in the energy, and therefore the frequency of the scattered light.

In the classical view regarding light as an electromagnetic wave, the Raman effect can be explained through the dipole moment theory. Since the molecular polarizability is a function of the vibrational coordinate, the generation of a dipole moment in a molecule by the incoming electric field is modulated by the molecular vibrations. Therefore, the induced dipole oscillates not only at the frequency of the incoming field, but also at the "side bands" at the frequencies between the excitation light and the molecular vibrations [217]. This results in the appearance of Stokes and anti-Stokes shift in the radiated light.

The left-hand side of Figure 2.20 shows the schematic of Raman scattering in a molecular energy level diagram. The interaction of a photon with a molecule in the ground state ($v = 0$ in S_0) may momentarily raise the molecule to a higher energy level, or a virtual state, which is not stable at room temperature and has a lifetime on the order of $\sim 10^{-15}$ s [218]. If the molecule leaves this unstable level, it can scatter or interact with a photon and return to the ground state. In this case, the scattered photon has the same energy as the exciting photon ($h\nu_L$) and Rayleigh scattering occurs. On the other hand, if the molecule falls back to an excited vibrational state, such as $v = 1$, the scattered photon's energy is now $h\nu_S = h\nu_L - h\Delta\nu$, where $h\Delta\nu$ is the difference in energy between the $v = 1$ and $v = 0$ levels. The frequency of the scattered photon thus is lower than that of the incident light, giving rise to what is called a Stokes line. Anti-Stokes line is observed when a molecule in an excited level ($v = 1$) rises to a higher, unstable level by interacting with an incident photon and then returns to the ground state upon scattering a photon. The scattered photon has a higher energy of $h\nu_{AS} = h\nu_L + h\Delta\nu$.

In either case one vibrational quantum of energy is gained or lost, so the the Stokes and anti-Stokes lines are equally displaced from the Rayleigh line. According to the Boltzmann population distribution, the relationship

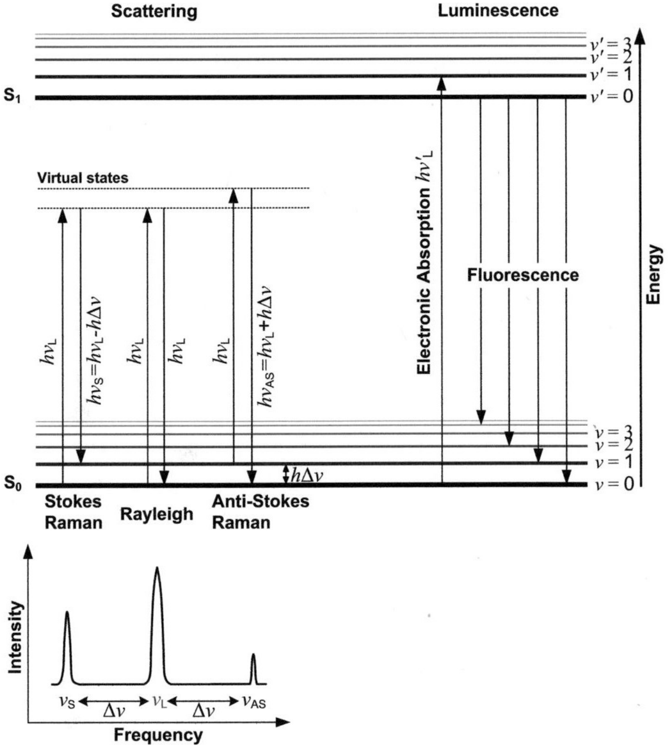

Figure 2.20: Example energy level diagram depicting, on the left-hand side, molecular Rayleigh and Raman scattering processes and, on the right-hand side, a photoluminescence process (fluorescence).

between Stokes and anti-Stokes intensities can be obtained as [219]

$$\frac{I_{\mathrm{S}}}{I_{\mathrm{AS}}} \approx \exp\left(\frac{E}{k_B T}\right) \tag{2.37}$$

where $k_B T$ is the product of Boltzmann's constant and thermodynamic temperature, E the vibrational energy. The probability of a molecule being in an excited state is exponentially lower than being in the ground state.

Therefore the anti-Stokes scattering occurs much less frequently than the Stokes scattering. This disparity can be reduced under increased temperature and it also depends on the spacing of the energy levels. For the less widely spaced rotational levels, the Stokes and anti-Stokes scattering are of similar magnitude. For vibrational levels which are spaced further apart, the anti-Stokes signal is significantly weaker than the Stokes signal. In most cases, the Raman photons collected and analyzed are the Stokes photons. Although the rarity of anti-Stokes photons results in much weaker anti-Stokes lines, they are sometimes favored in analysis due to absence of fluorescence interference, which can be a big problem for detection of Stokes lines.

The Raman effect probes the vibrational levels of a molecule, which depend on the kinds of atoms as well as their bond strengths and arrangements in a specific molecule. That is the reason why a Raman spectrum is also called the structural "fingerprint" of a molecule. Note that in nonresonant Raman scattering excitation, i.e., the scattered photons are not in resonance with any real electronic transition of a molecule, the interaction with the molecule is due to scattering and not due to an absorption/emission process. Therefore Raman scattering is much different from fluorescence (see Figure 2.20 on the right-hand side). In fluorescence, the incoming photon is absorbed by the molecule and causes an electronic energy state change. A fluorescent photon is later released when the molecule relaxes back to a lower energy state (on the order of $\sim 10^{-9}$ s) whereas Raman scattered photon is released instantaneously ($\sim 10^{-15}$ s). When the incident radiation is at a frequency near the frequency of an electronic transition of the molecule of interest, it provides enough energy to excite the electrons to a higher electronic state. This is so-called *resonance Raman scattering* (RRS). The vibrational modes associated with this particular electronic transition greatly increase the Raman scattering intensity, e.g., the resonance with a π-π^* transition enhances stretching modes of the π-bonds involved with the transition, while the other modes remain unaffected. Raman scattering in general is a very weak effect. Typically the Raman scattering cross sections per molecule range from 10^{-31} to 10^{-25} cm^2 [218, 220], about $\sim 10^5$ times weaker than Rayleigh scattering. In the case of RRS, the Raman scattering effect can be enhanced by a factor of $\sim 10^2$–10^6, but it is still a much weaker process compared to fluorescence, which has a luminescence cross section of 10^{-16} cm^2.

The overall Raman signal power depends on the Raman cross section σ_R, the excitation intensity $I(\nu_L)$ and the number of molecules in the probed volume N. The Stokes Raman signal $P^{RS}(\nu_L)$ can be described as [221]

$$P^{RS}(\nu_S) = N\sigma_R I(\nu_L) \tag{2.38}$$

The low Raman cross sections lead to very small or undetectable Raman signals. Various ways of sample preparation, sample illumination or scattered light detection have been invented for enhancing the Raman signal intensity. Advanced spectroscopy techniques, such as *stimulated Raman scattering*, *coherent anti-Stokes Raman spectroscopy* (CARS), *surface-enhanced Raman spectroscopy* (SERS) and *tip-enhanced Raman spectroscopy* (TERS) have contributed to the development of Raman spectroscopy and moreover broadened up its application fields.

Because the virtual states are not stationary, Raman (and also Rayleigh) scattering will occur in a way regardless of the incoming light frequency. The absolute frequency of scattered light changes with the excitation frequency, but the frequency shift, i.e., the quantized energy increment/decrement between excitation and scattered light ($h\widetilde{\nu}$) depends only on the differences in rotational and vibrational levels of the molecule and remains constant for one specific molecule:

$$\widetilde{\nu} = \frac{1}{\lambda_L} - \frac{1}{\lambda_{RS}} \tag{2.39}$$

where λ_L is the wavelength of the excitation source in unit of cm, λ_{RS} is the wavelength of one Raman line in unit of cm and $\widetilde{\nu}$ is the Raman shift in unit of cm^{-1}.

This independence on excitation frequency gives experimentalists the freedom to choose an assortment of lasers that are suitable for Raman scattering applications. However, the choice is not easy because several complicating factors and associated "trading rules" should be taken into account, which will be discussed in detail below.

2.4.2 Laser Raman spectroscopy

Laser excitation sources

In modern Raman spectroscopy, lasers are used as photon sources due to their highly monochromatic nature and high beam fluxes [5, 217, 222]. Among various commercially available lasers, the choice of one proper laser excitation source in Raman spectroscopy is not straightforward. First, it is important to consider the appropriate laser excitation wavelength. Since the scattering depends on the fourth power of the frequency [218], a blue/green visible laser or a UV laser results in an increase in scattering intensity by a factor of 15 or more when compared with infrared lasers. However, in addition to the aspect of sensitivity, the sample behavior must be considered for a compromise. The blue/green lasers often suffer the fluorescence problem. UV excitation leads in addition to a high risk of sample degradation and the Raman spectra may be rather different from normal Raman spectra due to the resonance with electronic transitions. Generally, blue or green lasers can be good excitation sources for inorganic materials and resonance Raman experiments (e.g., for carbon nanotubes and other carbon materials). Red or near infrared (660–830 nm) lasers are good for fluorescence suppression and ultraviolet lasers for resonance Raman on bio-molecules (such as proteins, DNA, and RNA) [222].

Additionally, the following basic requirements on laser sources which must be met for Raman spectroscopy: adequate laser power, high wavelength stability, high spatial mode quality and moderately narrow line width [5].

For specific applications of Raman spectroscopy, some novel laser excitation sources are developed and applied in Raman detections. For instance, pulsed lasers are typically utilized in the study of short-lived species. And in time resolved resonance Raman spectroscopy, pairs of laser pulses with different wavelengths are required to photolyse and then to Raman probe the transient species of interest [223]. A tunable laser is often employed as the excitation source to tune wavelength to conform to a specific resonance frequency and avoid fluorescence or thermal emission backgrounds [5, 220]. The broad spectral range of organic semiconductors and simple thin film deposition techniques are making organic semiconductor lasers very promising as truly tunable laser excitation sources. Before author's work, the only organic laser applied in Raman spectroscopy is the liquid dye laser.

Confocal Raman microscope

A Raman system typically consists of four major components: an excitation source (usually a laser), the sample illumination system, the wavelength selector (filter and spectrometer) and a detector (CCD camera, photo diode array or photomultiplier tube (PMT)). Generally, the excitation laser beam is focused through the microscope to create a micro-spot with a diameter on the order of 0.5–10 μm. The Raman signal from the sample is collected from a similar area, passes back through the microscope into the spectrometer and is there analyzed for spectral information.

Since the Raman spectroscopy is oftentimes dealing with selected portions of analytes or single molecule/cell, an outstanding resolution in three dimensions is required. A simple optical microscope assists in giving lateral (*XY*) spatial resolution, but does not give depth (*Z*) spatial resolution. For this purpose confocal optics were developed [224].

The three dimensional resolution is attained in such a way that all the light which is not coming from an in-focus plane is being blocked or removed afterwards. The way to achieve this is by putting pinholes in the excitation and detection paths. By setting a pinhole in front of the light source as an entrance aperture (as depicted in Figure 2.21(a)), the small pinhole aperture is refocused by the microscope objective. The second pinhole is placed between the microscope objective and the image plane as a detection aperture, which can reject the residual scattered rays originated from any out-of-focus points on the sample and functions as a spatial filter, as shown in Figure 2.21(b) and 2.21(c). Since scattered light from all other parts of the sample except the excitation focus spot is thus efficiently suppressed, the total amount of background scattered light is reduced by a few orders of magnitude without affecting the focal brightness.

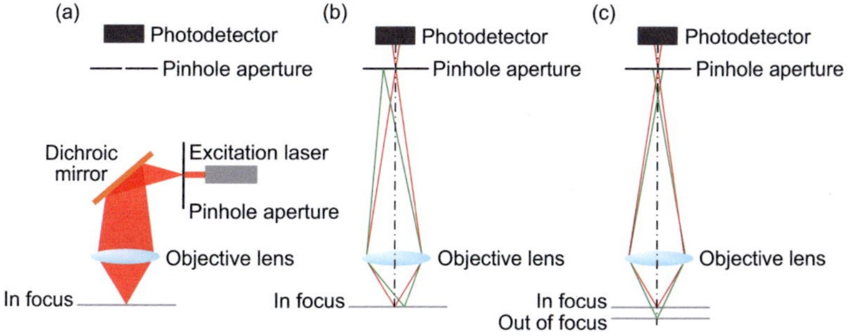

Figure 2.21: (a) The entrance pinhole aperture allows an illumination of a single spot on a sample. Spatial filtering with the confocal pinhole aperture of the horizontal (b) and vertical (c) out-of-focus rays.

The achievable lateral spatial (*XY*) resolution is primarily defined by the laser wavelength and the microscope objective. The theoretical diffraction limited spatial resolution, according to the radius of the Airy disc, is defined by the following equation [225]

$$\Delta x = \frac{0.61\lambda}{NA} \tag{2.40}$$

where λ is the wavelength of the laser, and *NA* is the numerical aperture of the microscope objective and is defined as follows:

$$NA = n \cdot \sin\alpha \tag{2.41}$$

In which n is the refractive index of the medium immersed between lens and sample and α is half the angle subtended by the lens at its focus. The concept of numerical aperture is closely related to that of the f-number. In a simple lens the f-number is the ratio between the lens' focal length f and its aperture D (effective diameter). The f-number is given by $N = f/D$. The relation between f-number and NA is:

$$NA = n \cdot \sin\alpha = n\frac{D}{2f} = \frac{n}{2N} \tag{2.42}$$

The depth (Z) spatial resolution (also referred to as axial resolution) depends strongly on the confocal design of the Raman microscope. A variety of models are presented in the literature to calculate the depth spatial resolution for various microscope configurations. Those that are most applicable to Raman spectroscopy are similar in form to the models in evaluating the depth of field: the depth spatial resolution (Δz, also called *focal depth*) is proportional to wavelength (λ) and refractive index (n) of the specimen medium, and inversely proportional to the square of the numerical aperture (NA). One equation commonly used to describe depth spatial resolution for the confocal configuration is given below (with the same variables as above) [226]:

$$\Delta z = \frac{2 \cdot n \cdot \lambda}{NA^2} \tag{2.43}$$

A smaller pinhole results in a smaller detection volume, and thus a lower background noise, but also fewer light to the detector. In practice, the pinhole should have approximately the same diameter as the FWHM of the Airy diffraction pattern generated by the lens at the pinhole's position (intermediate image plane). For a modular Raman system, optical fibers are usually applied in the confocal microscope to replace the pinholes. Commonly, a single-mode optical fiber is used to couple a laser beam into the microscope as an entrance aperture. More elegant than a conventional pinhole configuration, the output of the single-mode fiber can be considered to be equivalent to an ideal TEM_{00} laser beam diverging from its focusing point [224, 227]. Single-mode, few-mode and multi-mode fibers can also act as pinhole apertures in the detection path. The effective pinhole size can be controlled by using different fibers ranging from a single-mode fiber up to a large-core multi-mode fiber [224].

2.4.3 Surface-enhanced Raman spectroscopy

Since the discovery of the Raman effect, scientists have been constantly in search of proper ways to improve the Raman detection efficiency. In 1974, M. Fleischmann et al. discovered an unexpectedly strong Raman signal from pyridine on a rough silver electrode [228]. Afterwards, this result was confirmed in 1977 by D. Jeanmaire and R. V. Duyne [229], and independently by M. G. Albrecht and A. Creighton [230]. It was shown that the strong Raman signal was caused by a true enhancement of the Raman scattering efficiency and not by a scattering of more molecules. Within the next few years, strongly enhanced Raman signals were further verified in many different molecules attached to various rough metal surfaces and this effect was named ***S**urface-**E**nhanced **R**aman **S**cattering* (SERS). A simple illustration reflecting the concept of SERS nanostructures is shown in Figure 2.22(a).

The exact SERS mechanism is still in debate in the literature. Until now, two contributions are approved by the great majority of spectroscopists: One is the electromagnetic field enhancement contributed by *localized surface plasmon resonance* (LSPR) of metallic nanostructures and the other is the chemical/electronic enhancement based on electronic coupling between molecule and metal.

A simple model can help to understand the concept of electromagnetic SERS enhancement. As shown in Figure 2.22(b), the metallic nanostructure is illustrated as a small sphere with the complex dielectric constant $\varepsilon(\nu) = \varepsilon'(\nu) + i\varepsilon''(\nu)$ for a certain frequency ν in a surrounding medium

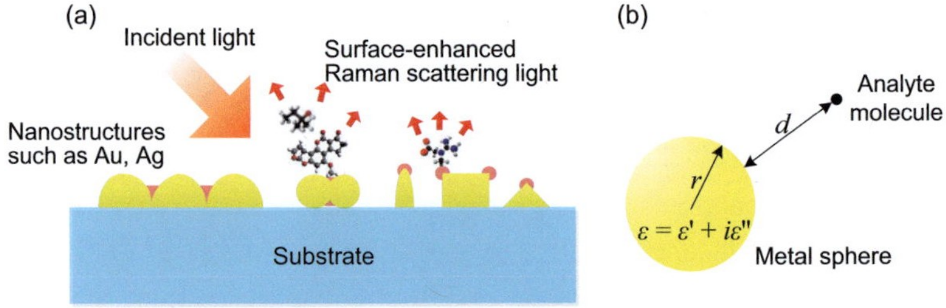

Figure 2.22: (a) Concept image of surface-enhanced Raman scattering. The Raman scattering of the adsorbed molecules is enhanced by the super-fine structures down to nanometer to micrometer level on metal surfaces such as gold and silver (adapted from [231]). (b) Simple schematic diagram for understanding the concept of electromagnetic SERS.

with a dielectric constant ε_0. A molecule in the vicinity of the sphere separated with a distance of d is exposed to a field E_M, which is a superposition of the incoming field E_0 and the field of a dipole E_SP induced in the metal sphere. The field enhancement factor $g(\nu)$ is the ratio of the field at the position of the molecule and the incoming field:

$$g(\nu) = \frac{E_\mathrm{M}(\nu)}{E_0(\nu)} \approx \frac{\varepsilon(\nu)-\varepsilon_0}{\varepsilon(\nu)+2\varepsilon_0}(\frac{r}{r+d})^3 \tag{2.44}$$

The resonant excitation conditions of metal sphere surface plasmons comprise the real part of $\varepsilon(\nu)$ equaling to $-2\varepsilon_0$ and a minor imaginary part of $\varepsilon(\nu)$.

The SERS enhancement is particularly strong when both laser $g(\nu_\mathrm{L})$ and scattered fields $g(\nu_\mathrm{S})$ are in resonance with the surface plasmons. Taking into account enhancement for the laser and the Stokes field, the electromagnetic enhancement factor for the Stokes signal power can be written as [221]

$$|g(\nu_\mathrm{L})|^2|g(\nu_\mathrm{S})|^2 \approx |\frac{\varepsilon(\nu_L)-\varepsilon_0}{\varepsilon(\nu_L)+2\varepsilon_0}|^2|\frac{\varepsilon(\nu_S)-\varepsilon_0}{\varepsilon(\nu_S)+2\varepsilon_0}|^2(\frac{r}{r+d})^{12} \tag{2.45}$$

The electromagnetic SERS enhancement does not require direct contact between molecule and metallic nanostructure, but it decreases rapidly with growing distance due to the dipole field decay factor $(1/d)^{12}$. In the situation of a small Stokes shift, $g(\nu_{\mathrm{L}})$ and $g(\nu_{\mathrm{S}})$ are at approximately the same frequency and the electromagnetic enhancement scales with the fourth power of the local field at the metallic nanostructure g^4 [221, 222, 232]. Although experiments were proposed at the inception of SERS to measure this effect, studies performed were inconclusive owing to the limitations in instrumentation and poor definition of SERS substrates. The ideal experiment is to have a continuously tunable excitation laser and detection scheme over the bandwidth of a well-defined LSPR. The electromagnetic enhancement for isolated single silver and gold colloids is on the order of $\sim 10^6$–10^7 [233], which can be increased up to nearly 10^{11} for sharp features, e.g., the midpoint between two silver or gold spherical particles separated by a gap of 1 nm [234, 235]. An extremely large SERS electromagnetic enhancement up to 10^{12} was theoretically predicted for the strongly confined so-called "hot spot" regions [221].

The overall SERS signal power can also be described with a formula similar to that for "normal" Raman processes in Eq. 2.38 [221]:

$$P^{\mathrm{SERS}}(\nu_{\mathrm{S}}) = N'\sigma^{\mathrm{R}}_{\mathrm{ads}}|g(\nu_{\mathrm{L}})|^2|g(\nu_{\mathrm{S}})|^2 I(\nu_{\mathrm{L}}) \qquad (2.46)$$

where N' is the number of molecules which are involved in the SERS process and may be smaller than the number of molecules in the probed volume N and $\sigma^{R}_{\mathrm{ads}}$ denotes the increased cross section for the new Raman process of the adsorbed molecule.

It indicates that the overall enhancement is not only due to the electromagnetic enhancement, but also partly relates to the increased cross section $\sigma^{R}_{\mathrm{ads}}$. The experimental observations also provide a strong molecular selectivity and a dependence of SERS intensity on the chemical nature of the molecule. Moreover, the predicted electromagnetic SERS enhancement factors leaves a gap of about two orders of magnitude to the experimentally observed SERS enhancement factors [232, 233], which suggested the existence of additional enhancement contribution accounting for the missing factors.

In general, the chemical SERS enhancement factor is considered to contribute to the overall enhancement factors on the order of $10–10^3$ as arising from the excitation of adsorbate localized electronic resonances or metal-to-adsorbate charge-transfer resonances [232]. Clearly, chemical enhancement should occur only from molecules directly attached to the surface (characterized by an Ångström length) and consequently should increase only up to monolayer coverage.

The vibrational frequencies and Raman scattering cross sections of SERS spectra are generally different from those measured in neat samples [236]. As a consequence, additional bands may become observable in SERS spectra, or existing bands may be attenuated beyond the detection limit. These modification effects in SERS spectra are still under discussion. Several combined theoretical and experimental studies have been reported that they can be explained by surface selection rules due to the effect of molecule orientation on SERS substrates [237, 238]. The vibrations, which orient perpendicular to the surface, may experience the highest enhancement [239]. Another hypothesis attributes the selective enhancement for different Raman modes to the changes in molecular electronic structure of the analyte [240]. Both arguments are related to the chemical enhancement mechanism [241].

Although the theoretical analysis is useful to estimate the SERS enhancement, it is often simpler to experimentally measure the SERS enhancement factor (EF), which can be described by [232]

$$\mathrm{EF} = \frac{I_{\mathrm{SERS}}/N_{\mathrm{surf}}}{I_{\mathrm{ref}}/N_{\mathrm{vol}}} \tag{2.47}$$

where I_{SERS} is the SERS intensity, N_{surf} is the number of molecules within the "hot spot" regions under laser excitation, contributing to the measured SERS signal, I_{ref} is the non-SERS intensity from the reference sample, and

N_{vol} is the number of molecules in the laser excitation/detection reference volume. I_{SERS} and I_{ref} can be independently measured for a given molecule. N_{surf} and N_{vol} must be carefully handled by evaluating the spot size and the probe volume.

3 Laser Fabrication and Characterization Techniques

This chapter introduces various fabrication and characterization techniques for organic semiconductor lasers used in this work. The corrugation grating substrates are fabricated by electron beam lithography, nanoimprint lithography and interference lithography. Thin films of organic semiconductors are deposited on nanopatterned substrates by sublimation in high vacuum or by liquid processing from solutions.

3.1 Fabrication of DFB gratings

3.1.1 Electron beam lithography

Electron beam lithography (EBL) is a technique commonly used for the fabrication of micro- and nanostructures, based on the chemical modification of polymeric resist films caused by electron irradiation. The main advantages of EBL over conventional photolithography techniques include high resolution and versatile pattern formation. In this work, this method was applied to fabricate the silica gratings for organic semiconductor lasers and the nanopatterned silicon master templates for nanoimprint lithography.

In this work, the nanostructures are created by using the EBL system (Vistec Lithography, Vistec VB6) at the Institute of Microstructure Technology (IMT-KIT). Before the electron beam (e-beam) writing process, a resist, in this work a 50–80 nm thin layer of poly(methyl methacrylate) (PMMA) 950k A2 (MicroChem), is spin-coated onto a thermally oxidized silicon wafer (for DFB gratings) or a silicon wafer (for nanopatterned templates). After focusing with a magnetic lens, the electron beam scans the sample and exposes the resist. Upon exposure to high energy, the resist molecules are modified, resulting in a changed solubility. Note that the nanostructures on the corresponding template/stamp should be inverted patterns in comparison to the target replicated samples. Therefore a proper resist and layout design must be prepared prior to the e-beam writing process. EBL usually allows a very small beam size and thus ensures a nanopatterning with a very high resolution down to sub-10 nm [242]. However, due to the serial writing principle, EBL is time-consuming and therefore has a low throughput.

After e-beam writing, a developing process removes the corresponding exposed regions where the local resist becomes soluble (for positive resist). The remaining patterned resist layer serves as a mask for a later lift-off or etching process. Finally, the desired applicable patterns can be obtained after removing the resist or metallic mask in the lift-off process. Figure 3.1 shows the schematic illustration of the fabrication steps of SiO_2 surface gratings on a thermally oxidized silicon wafer and nanostructures on the silicon master templates.

For combined microstructures (waveguides or trenches for nanostructures), an additional optical UV-lithography will follow the EBL and corresponding etching process. To realize this, the positive photoresist AZ 1505 (AZ Electronic Materials) is spin-coated onto the substrate with a thickness of 0.5 µm. Upon UV exposure with a dose of 50 mJ cm^{-2} under a mask aligner (EVG 620), which can be approached by adjusting the photomask with the EBL-created label, the microstructures can be defined onto the sample. The steps of 4–6 in Figure 3.1(b) show the subsequent fabrication steps of microstructures after nanograting etching.

In this work, the final nano- and microstructures for the silicon templates are generated by SF_6-based reactive-ion etching. To ensure an optimum anisotropic etching with smooth straight side walls, the cryogenic etching process was performed at -125 °C to transfer the nanopatterns into the silicon substrates. The etching process parameters are listed in Appendix A.1. The etching period depends on the etching depth of the desired structures and amounts to 3–100 s.

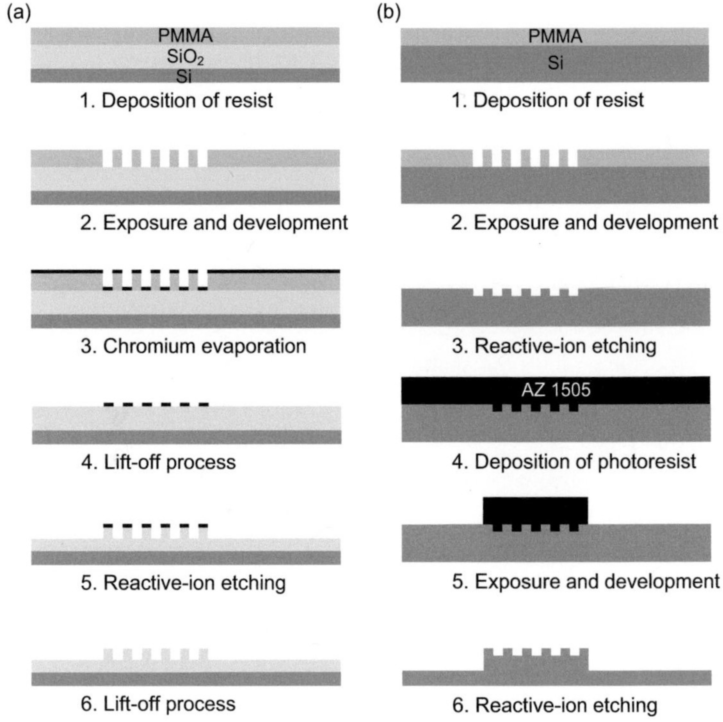

Figure 3.1: Schematic illustration of fabrication processes of (a) SiO_2 gratings on a thermally oxidized wafer and (b) silicon master templates as replication stamps for nanoimprint lithography.

3.1.2 Nanoimprint lithography

Nanoimprint lithography (NIL) is an alternative method for the fabrication of nanometer-scaled patterns. In this technique, a rigid stamp (or master, template, shim) is needed to create patterns by mechanical deformation of polymer substrates or polymer layers on rigid substrates. The process has the advantages of potential low cost, high throughput and high resolution fabrication. NIL promises a mass-production of extremely fine structures with feature sizes down to 10 nm [243]. The resolution of NIL can be further improved by the fabrication of a rigid mold with sub-10 nm features [244]. Today, several NIL techniques exist, e.g., UV-assisted NIL, thermal NIL (also known as *hot embossing*), soft lithography, roll-to-roll imprinting, etc. Two common techniques are UV-assisted NIL and hot embossing.

In UV-assisted NIL, stamps, often referred to as templates, are imprinted into UV-curable polymers at room temperature or slightly above. When the stamp is filled with a UV-curable polymer, the polymer is exposed to UV radiation directly or behind the stamp. Subsequently, the stamp is removed. This procedure allows the spatially localized imprint through photomasks on certain area of one substrate. It also allows large-area patterning by a step-and-repeat process on the whole substrate [245]. These make UV-assisted NIL a suitable candidate to fabricate DFB-corrugations for organic semiconductor lasers and nanostructures for surface-enhanced Raman spectroscopy. However, it is mandatory to use specific photoactive materials in the UV-NIL process. This limitation severely constrains the fabrication of lab-on-a-chip (LOC) systems [246]. Instead, another technique hot embossing is employed to fabricate the polymer grating substrates in this work.

Hot embossing

In hot embossing, a thermoplastic polymer is deformed by pressing the stamp into the polymer at a temperature above the polymer's glass-transition temperature (T_g). Hereafter, the polymer is cooled down below the glass-transition temperature and the structures are preserved [245]. In this work, a 4-inch patterned n-doped silicon $\langle 100 \rangle$ wafer is used as a template. Poly(methyl methacrylate) (PMMA) and cyclic olefin copolymer (COC) are applied as the thermoplastic polymer substrate, which are often used in LOC devices.

The replication of nanostructures is performed with the hot embossing equipment at IMT (Jenoptik, WUM02), where the silicon template and the raw polymer substrate are placed between two plane-parallel heating plates in a vacuum chamber. Before replication, a monolayer of perfluorodecyltrichlorosilane (FDTS) is deposited as an anti-adhesion layer onto the silicon template using *chemical vapor deposition* (CVD) (Applied MicroStructures, MVD100E) for easy release [247]. An FDTS monolayer has at one end the reactive group -$Si_{(teflon)}Cl_x$, which allows bonding onto surfaces terminated with hydroxyl (-OH) groups, e.g., $Si_{(surface)}$-OH groups of the native oxide on Si surfaces, to form a chemical bond such as -$Si_{(teflon)}$-O-$Si_{(surface)}$. At the other end, it has typically flourinated organosilanes with a teflon-like tail consisting of -$(CF_2)_xCF_3$ which reduces surface energy and prevents the silicon wafer from sticking to the polymer surface.

The polymer substrate can be put into the middle of a silicon template and an unstructured silicon wafer with an anti-adhesion layer. The practical experience indicates that adding a thin graphite layer can help to compensate the unevenness during the embossing process [247]. To avoid possible dust and dirt from the graphite layer, it can be placed next to the silicon template/unstructured silicon wafer between two aluminum foils.

Hot embossing is performed in vacuum ($\sim$ 15 Pa). It has three main working periods: heating up to a sufficiently high temperature, applying mechanical pressure and cooling down to ambient temperature.

The heating procedure takes 20–30 min, depending on the employed substrate materials. Principally, the achieved temperature should by some degree exceed the glass transition temperature of the polymer. For an optimum replicated polymer structure consistent with the original template, a high mechanical force and a small cooling rate should be applied in the embossing process. However, since the silicon template is rather fragile, no pressure higher than 15 kN is applied in this work. As a compensation, a higher heating temperature should be provided for a reasonable result. The mechanical pressure under the target temperature is applied for 10–15 min. In order to eliminate the difference in expansion/shrinking caused by thermal-expansion mismatch between polymer and template, the mechanical force is held until the polymer is cooled to a certain temperature below 50 °C. The overall hot embossing process lasts for 45–60 min.

The process parameters for polymer materials PMMA and COC are listed in Table 3.1. The glass transition temperature of these two materials are $T_g \approx 105$ °C for Hesa-Glas VOS PMMA and $T_g \approx 138$ °C for Topas 6013 COC [247].

Parameters / Materials	Thickness (µm)	Temperature (°C)	Embossing time (min)	Pressure (MPa)
Hesa-Glas VOS PMMA	1024	190	15	2.7
Topas 6013 COC	102	190	15	2.6
Topas 6013 COC	254	190	15	3.3

Table 3.1: Replication parameters of the hot embossing processes for PMMA and COC substrates.

Laser-assisted replication

Two main problems of hot embossing should be noted. One is the difficulty in fabricating pre-defined partial structures from a rigid master stamp with various structures for different purposes. Another issue is the long heating and cooling time affecting the throughput. Both can be detrimental especially for the already fabricated LOC passive photonic elements and microfluidic channels. Such problems can be solved by laser-assisted replication using an infrared diode laser beam [248, 249, 250, 251].

In *laser-assisted replication*, a rigid mold, e.g., a silicon master stamp is pressed under a polymer substrate and the contact area is subsequently heated up by a near-infrared laser beam from above, as shown in Fig. 3.2(a). Since the used polymeric materials, PMMA and COC, are transparent to the applied laser wavelength, the laser beam is mainly absorbed by the silicon surface in contact with the polymer substrate. Due to heat conduction the temperature of the polymer, which is in direct contact with the master stamp, increases. When the temperature exceeds the glass-transition temperature of the polymer, the nanostructures on the rigid mold are replicated into the polymer substrate. After cooling down to ambient temperature, the polymer substrate is detached from the rigid mold while the nanostructures are preserved.

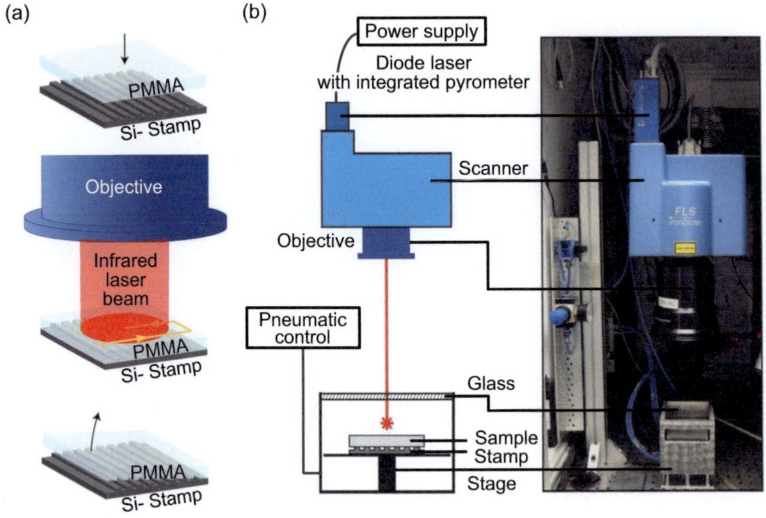

Figure 3.2: (a) Scheme of the laser-assisted replication process: positioning of the PMMA substrate on a silicon master stamp (top), execution of the replication (middle) and detachment of the sample (bottom). (b) Schematic illustration (left) and photograph (right) of the applied laser system.

Instead of large-area heating, this method enables a selective local replication and brings the unique benefit to replicate only parts of the existing structures from a master stamp. Additionally, since only the pre-defined areas are to be heated, it allows a subsequent additive replication while all other areas stay uninfluenced.

To perform replications, a high-power diode laser system (Fisba Optik, FLS Iron High Power Diode) with an emission wavelength of 940 nm and a maximal power of 50 W is used, which locates at the Institute for Applied Materials – Applied Materials Physics (IAM-AWP). As shown in Figure 3.2(b), the diode laser and a pyrometer (Fisba Optik, PyroS), for in-situ monitoring of the temperature at the laser focus plane, are integrated within the laser scanner unit. The polymer substrate and the silicon master stamp are fixed in a pneumatic stage, applying a constant pressure. There are two basic operational modes of the system: constant laser power or constant heating temperature. In this case, the temperature regulation is chosen in order to achieve a homogeneous temperature distribution at the contact surface. With the help of a visible laser beam at 670 nm, the path and the working range of the infrared laser could be verified before starting the replication process. The *beam parameter products* (BPPs) of the applied diode laser are 15 mm×mrad and 30 mm×mrad in two orthogonal directions. Utilizing an F-Theta objective lens (Fisba Optik) with a focal length of 163 mm, the size of the elliptical laser focus spot on the work plane is 0.55 mm × 0.7 mm. Attributed to the applied scanner system, the maximum laser processing area can reach up to 100 mm × 100 mm.

The replication results and its quality can be influenced by various factors, e.g., the local heating temperature T, the diode current I and the laser output power P, the laser writing velocity v, the applied pressure on samples p, the line offset (writing distance between two subsequent lines) Δx, the writing repetition times N and so on [252].

After replication, the clamping pressure is released and the PMMA substrate is cooled down to ambient temperature. Since there is no large-area heating in the laser-assisted replication process, the cooling time as well as the whole replication period is much shorter as compared to the hot embossing process. After cooling down, the polymer substrate with replicated nanostructures is detached from the silicon template and the process can be repeated with a fresh COC substrate. One should note, however, that the laser focusing and scanning parameters have to be optimized in order to avoid a destruction of the stamps because of too large thermal gradients in silicon. Even there is an anti-adhesion layer deposited on the silicon template, the separation from the polymer sample is possibly not spontaneous. It can be achieved with the help of nitrogen spray or a razor-sharp blade. It is also helpful to use drops of ethanol to release the silicon template. Since the laser-assisted replication is undergoing in ambient atmosphere, the residue air may cause a small gap stopping the amorphous polymer filling into the template trenches. It has been observed that in the replication process of high-density nanostructures, the height of the patterned polymer features is about 75 % of the original template feature depth [252].

The polymer substrates used in this process consist of PMMA (Evonik, Plexiglas® XT) and COC (TOPAS COC 6013), whose glass transition temperatures are 103 °C and 138 °C, respectively. This novel replication method was applied in the fabrication of one-dimensional grating substrates for organic semiconductor lasers and efficient nanopillar arrays as active SERS substrate. The fabrication details and corresponding applications will be discussed in detail in Section 4.3.

3.1.3 Interference lithography

Interference lithography (IL) (also known as *holographic lithography*) is a technique for patterning large-area periodic arrays of fine features, without the usage of complex optical systems or photomasks. The benefit of using IL is the quick generation of dense features over a wide area without time-consuming mechanical scanning [253]. It is especially suitable to fabricate large-area DFB-corrugation substrates for organic semiconductor laser applications.

The principle of IL is similar to that of interferometry or holography. The superposition of two or more electromagnetic waves of the same wavelength with a constant phase relation leads to a periodic modulation of the total intensity and thus interference occurs. This interference pattern comprises a periodic series of fringes representing intensity minimum and maximum. The general working principle of two-beam laser interference lithography used in this work is depicted in Figure 3.3.

The IL-setup at the Light Technology Institute (LTI-KIT) employs a solid-state deep ultraviolet (DUV) laser (Crylas) with an operation wavelength of 266 nm, and a coherence length > 1000 m. After splitting the laser beam into two, both beams are expanded and directed towards the center of a photoresist-coated sample using rotatable concave mirrors. It is assumed

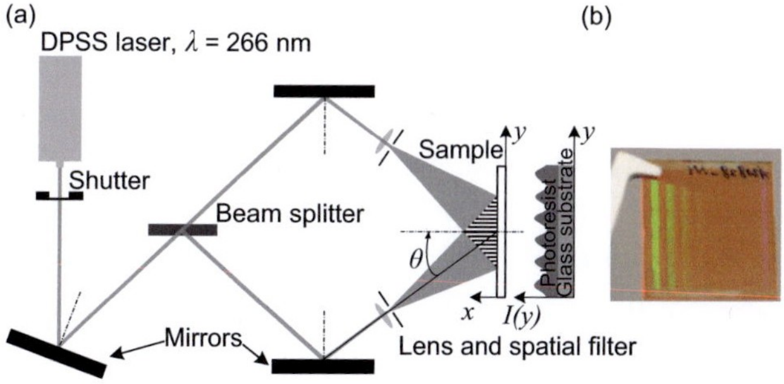

Figure 3.3: (a) Schematic illustration of laser interference lithography setup. (b) Photograph of a fabricated sample.

that if the traveling distance to the sample holder is sufficiently large, the light fields of the two light cones reach the sample surface as plane waves, where they overlap and interfere. In the beam overlapping area a characteristic sinusoidal bright-dark-interference pattern emerges and it serves for the exposure of the photoresist substrate. The two laser beams meet on the substrate's upper surface (represented here as plane waves), inclosing an angle of 2θ. The interference pattern period Λ can be expressed by the following formula:

$$\Lambda = \frac{\lambda}{2n\sin\theta} \tag{3.1}$$

where λ is the operation light wavelength and n is the refractive index of the top ambient medium, which is equal to 1 when the exposure is implemented in air. θ represents the angle of one beam from the substrate normal.

The patterning of one-dimensional (1D) sinusoidal corrugations with a periodicity of $\Lambda_x = 400$ nm is used for 1D-DFB lasers. Two-dimensional (2D) sinusoidal corrugations with a periodicity of $\Lambda_x (= \Lambda_y) = 400$ nm are applied for 2D-DFB lasers. The photoresist substrate is exposed twice by the UV-laser, one after another via rotating the sample by 90°.

After exposure, the samples are developed and the corresponding exposed/unexposed regions (according to the positive/negative photoresists) can be removed. In this work, two negative photoresists are used in the IL process to fabricate the grating substrates: SX AR-N 4800/16 (Allresist) and SU-8 2005 (MicroChem). The appropriate exposure and development parameters can be found in Appendix A.2. The patterned photoresist structures can serve as a basic mask for the later lift-off and etching process. However, in this work the patterned one-dimensional and two-dimensional sinusoidal gratings with a period of $\Lambda_x (= \Lambda_y) = 400$ nm are directly utilized as grating substrates for Alq_3:DCM lasers.

3.2 Thin film deposition

As introduced in Section 2.1, the organic semiconductor materials are commonly classified into two categories: small molecules and conjugated polymers. Small molecule materials are typically deposited by thermal evaporation in vacuum, whereas polymers are handled with solution processes. In this section, thermal evaporation is introduced first and then solution processing methods are discussed with an emphasis on horizontal dipping.

3.2.1 Thermal evaporation

Thermal evaporation (also known as *sublimation*) is an established technique to deposit thin films of metals, dielectrics or organic small molecules. The process takes place in a vacuum chamber ($\sim 10^{-4} - 10^{-5}$ Pa). The target materials, which are placed in crucibles at the bottom of the chamber, are heated until sublimation is observed on the samples placed on top of the chamber, as depicted in Figure 3.4(a). The evaporation speed depends on the temperature of the crucible and the deposition rate can be monitored in real-time by a quartz crystal microbalance (QCM). Very accurate thickness control can be achieved and multilayer deposition is easy to perform. The deposition of doped or composite materials, e.g., host-guest small molecule systems, can be realized by co-evaporation of different materials at different rates. Since this deposition technique can not be used for conjugated polymers and requires a costly vacuum chamber, intense efforts are devoted to simpler solution-based techniques. However, thermal evaporation

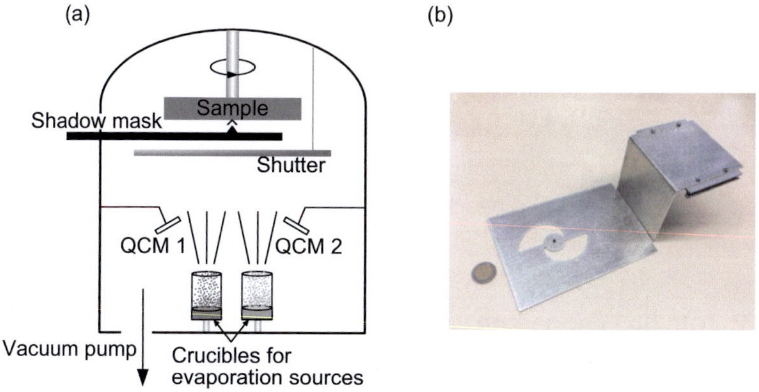

Figure 3.4: (a) Schematic diagram of a thermal evaporation system with two co-evaporation sources. (b) Photograph of a aluminum shadow mask used to fabricate a wedge-shaped organic active layer.

still dominants the OLED market owing to the outstanding deposition homogeneity, the possibility of spatially defined sublimation and multi-layer deposition for complex organic electronic devices.

In this work, the thermal evaporation process is used for fabricating Alq_3:DCM organic lasers. To eliminate the inhomogeneity in the deposited film thickness, a rotating sample holder is equipped in the evaporation apparatus at LTI (Kurt J. Lesker).

For the deposition on spatially defined sites on the target sample, e.g., integrated DFB laser pixels on a LOC system, a shadow mask with accurately defined opening features is required to protect the rest of the sample. The conventional steel shadow mask fabricated by laser cutting has disadvantages due to rough cutting edges and inaccurate square definitions. It is replaced with a nickel evaporation mask fabricated by electroplating. The fabrication detail can be found in Ref. [247]. Because the nickel mask has a small thickness of $\sim 50\,\mu\text{m}$, it can be flexibly arranged in contact with the target grating sample, which results in a sharp edge profile in the final functional lasers.

One shadow mask can be applied in the evaporation process to obtain a thickness gradient in the organic film, thus allowing an achievement of a continuously tunable organic semiconductor laser. By aligning the center of the openings coaxially to the sample holder, a radial film thickness gradient is created when the mask and sample holder rotate relatively to each other around a common axis during evaporation. The development of this technique were introduced in detail by Klinkhammer et al. [121]. In this work, an aluminum shadow mask with radially decreasing opening angle was used to achieve a linear decrease of Alq_3:DCM film thickness from the inner side to the outer side, as shown in Figure 3.4(b). The grating orientation of the substrate was aligned parallel to the resulting film thickness gradient in order to have a constant film thickness for the optical feedback [121].

3.2.2 Solution-based techniques

Deposition from solution is the main process used for organic conjugated polymers. Its simple operation and numerous fabrication varieties make it very promising for the fabrication of large-area and flexible organic semiconductor devices. The film thickness, optical properties and film homogeneity depend on the choice of the solvent and the concentration of the active material. The general fabrication steps consist of (1) preparation of a solution of the active medium in a proper organic solvent, (2) deposition of the solution onto a sample substrate by techniques such as *spin-coating*, *dip-coating* [254, 255], *doctor-blading* [256], *screen printing* [257, 258], *ink-jet printing* [259, 260, 261], *horizontal dipping* [22, 262, 263], etc. and (3) evaporation of the remaining solvent.

In this work, spin-coating, horizontal dipping and ink-jet printing are mainly used to fabricate the $F8_{0.9}BT_{0.1}$ and $F8_{0.9}BT_{0.1}$:MEH-PPV lasers. The newly introduced method of ink-jet printing of lasers will be discussed in detail in Section 4.2.

Spin-coating

As one of the simplest organic material deposition methods, spin-coating is commonly used in organic laser fabrication. To perform the deposition, an excess amount of polymer solution is dispensed onto the substrate, which is vacuum clamped for spinning. It is then accelerated up to the final desired rotation speed (usually 500–5000 rpm) in order to spread the fluid by centrifugal forces. Rotation is continued for about 20–50 s, with excess fluid being spun off the edges of the substrate, until the desired film thickness is achieved. The solvent is usually volatile, providing its simultaneous evaporation. Sometimes, a post spin-coating thermal treatment is applied to ensure complete evaporation of the solvent.

Very similar to spin-coating, or in some cases considered as the same process, *spin-casting* is also a method of utilizing centrifugal force to deposit organic thin films. The only difference from spin-coating is that it casts solution materials to form a disk mold shape.

In this work, spin-coating is mainly applied for the deposition of resists in the lithography process. Two main problems of this method are the low usage of the polymer solution and inhomogeneity for highly concentrated polymer solution.

Horizontal dipping

To get over the deficiencies of spin-coating, a deposition method, called horizontal dipping, is developed for the fabrication of organic semiconductor DFB lasers (DFB-OSLs). In horizontal dipping, a reservoir of coating fluid is maintained between the substrate and a horizontal metal rod (coating barrier) placed $\sim$ 1 mm above the substrate by capillary action. By drawing the barrier, a thin film is deposited upon the substrate, as illustrated in Figure 3.5(a). The resulted film thickness is accurately controlled by the properties of the active organic materials and the coating solvents as well as by the drawing speed of the coating barrier [262, 263, 264]. The transport of the barrier through a meniscus of the solution is similar to the dip-coating method, but different from the vertical withdrawing. The wet film is formed by transporting the barrier horizontally, so it is named "horizontal dipping".

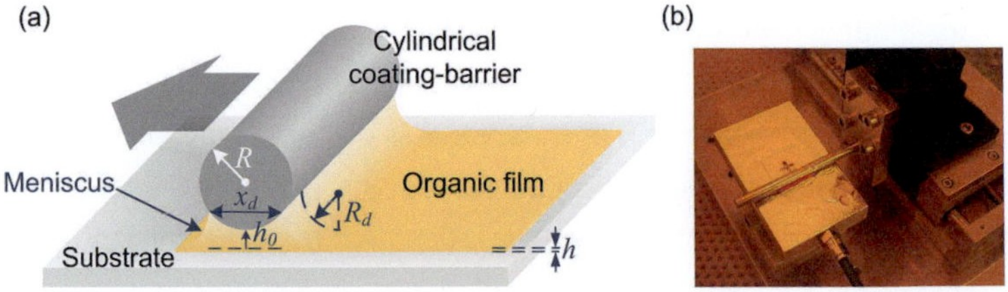

Figure 3.5: (a) Schematic illustration of horizontal dipping. (b) Photograph of the main setup of horizontal dipping.

As early as in 1942, Landau and Levich have found that by dragging a moving plate with the liquid, the left liquid film thickness can be determined by [265]

$$h = 1.34(\frac{\mu U}{\sigma})^{2/3} \cdot R_d \tag{3.2}$$

$$n \cdot R_d = (\frac{x_d^2}{2R} + 2h_0) - h \tag{3.3}$$

where μ and σ represent the viscosity and surface tension of the coating solution respectively, U the drawing speed, and R_d the radius of the curvature of the downstream meniscus. R, h_0 and x_d are the radius of the cylindrical coating barrier, the minimum gap height and the distance of meniscus edges (as depicted in Figure 3.5(a)) respectively. The contact angle is measured on the contact line at the interface between solution and coating barrier. For the coating processes, the contact angle is $0°$ and thus n is 2.

The horizontal-dipping apparatus is shown in Figure 3.5(b) and is an inhouse-assembled machine consisting of two main modules: a sample stage and a stainless steel coating barrier. The coating barrier is a cylindrical rod with a diameter of 6.3 mm. As the core component of the setup, it is transported by a stepper motor (EMIS, SM42051), which is connected to a controller (TRINAMIC Motion Control, TMCL-310).

Although the device construction of horizontal dipping is very similar to that of doctor-blading, they are based on distinctly different mechanisms. Doctor-blading is a metered coating process, whose film thickness is almost equal to the gap height. Even though a small perturbation is added to the final result due to the motion of fluid, there is no common explanation and estimation of this deviation. Horizontal dipping is a pre-metered coating process, whose film thickness can be calculated by a regular mathematical expression with capillary number $C = (\mu U/\sigma)$ and drawing speed (see Eq. 3.2). The initially formed downstream meniscus is the entire solution to be deposited and the quantity of solution can be estimated and measured. From the view of fabrication sequence, the first step of doctor-blading is to spread the coating solution onto the substrate and the film is newly shaped by transporting the blade. Therefore, large quantity of the organic solution is scraped off and wasted. Less viscous and volatile solutions are not suitable for doctor-blading. The first step of horizontal dipping is to place the coating barrier at the front edge of the sample, followed by introducing the coating solution into the gap, so it will automatically form a uniform meniscus of the solution. With horizontal dipping, 30–50 µl organic solution is sufficient for a deposition onto a 25×25 mm^2 grating sample.

Because the film thickness depends on the drawing speed of the coating barrier, it is possible to apply horizontal dipping to fabricate wedge-shaped organic films with controlled thickness. This can be achieved with an accelerated or decelerated drawing speed [22]. Using this simple technique, a solution-processed continuously tunable organic semiconductor laser can be easily built. The tunable spectral range can reach up to $\sim$ 38 nm using $F8_{0.9}BT_{0.1}$:MEH-PPV as the active medium [266]. In the fabrication, it is necessary to align the grating substrate so that the resulting thickness gradient is parallel to the grating orientation in order to provide uniform optical feedback in the region of interest. Recently, the wedge-shaped thin films deposited by horizontal dipping have been employed to study the spatial mapping of photocurrents in organic solar cells [267] and organic light emitting diodes [268]. In this work, horizontal dipping is used to deposit homogeneous conjugated polymer thin films on corrugated grating substrates.

3.3 Optical characterization technique

The optical characterization setup is shown in Figure 3.6. For the optical

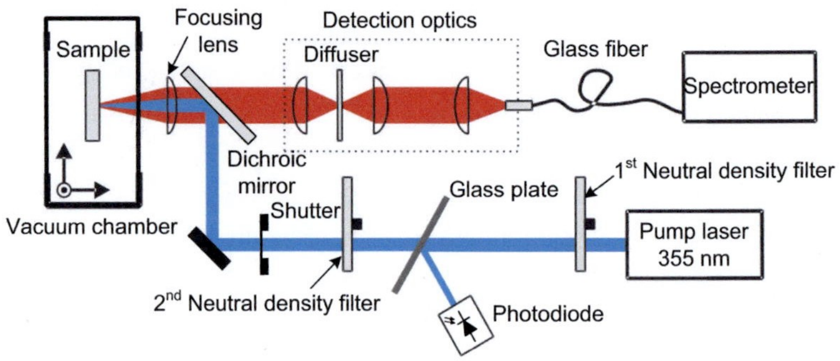

Figure 3.6: Schematic illustration of the calibrated optical characterization setup (adapted from [98]).

characterization of organic DFB lasers, the samples were optically excited by a diode-pumped, actively Q-switched frequency-tripled neodymium-doped yttrium orthovanadate (Nd:YVO_4) laser (AOT-YVO-20QSP) with a wavelength of 355 nm and a pulse length of about 0.5 ns. The repetition rate could be tuned up to 20 kHz. The pump pulse energy was adjusted with a variable neutral density filter and measured with a calibrated gallium arsenide phosphide photodiode connected to an oscilloscope (Tektronix, TDS2024C). The second neutral density filter was ready to use for manual change of intensity. The sample was kept in a vacuum chamber ($< 5 \times 10^{-3}$ Pa) to protect the active material from photooxidation. A converging lens with a focal length of 75 mm was used to adjust the excitation area of the pump spot by moving the vacuum chamber with the sample relative to the focal plane. Emission from the sample was collected using the converging lens for the pump beam, then directed through a dichroic mirror (AHF Analysentechnik, z355RDC) as well as detection optics and coupled into a multi-mode optical fiber. Further on, the emission spectra were analyzed by a spectrograph (Acton Research Corporation, SpectraPro 300i, variable gratings) connected to an intensified charge-coupled device (ICCD) camera (Princeton Research, PiMax 512). The vacuum chamber containing the sample could be moved in all three dimensions relative to the pump beam using a motorized precision stage. This allowed a spectrally and spatially resolved characterization of the lasers. The position-dependent dimensions of the laser spot on the sample were determined using the moving edge method and fitted with a Gaussian beam profile along the horizontal (x) and vertical (y) axis [269]. The beam emitted from the pump laser showed a slight elliptical shape. The diameters of the pump spot were taken as the extension of the pump spot along x and y for which the intensity was above $1/e^2$ of the intensity maximum.[1]

[1] In this work, the pump spot size is represented as $D_x \times D_y$, where D_x is the spot diameter in horizontal direction and D_y is the one in vertical direction.

4 Spatially Defined Organic Semiconductor Lasers for LOC Applications

The integration of organic semiconductor DFB lasers (DFB-OSLs) into all-polymer chips is promising for spectroscopic applications. This chapter describes how to realize spatially defined organic lasers ("laser pixel") on an all-polymer chip platform. Two technologies, local definition of the active material and laterally controlled fabrication of DFB-corrugations are developed. First, an overview of existing approaches is given to obtain a general state-of-the-art. A localized manipulation of conjugated polymer via ink-jet printing and a novel fabrication of DFB-grating pixels via laser-assisted replication are demonstrated in the following sections. A summary and discussion part will be given at the end of this chapter.[1]

4.1 State of the art

A lab-on-a-chip (LOC) device integrates one or several laboratory functions on a single chip of only square millimeters to a few square centimeters in size, which deals with extremely small fluid volumes down to less than picoliters [272]. The advantages of using LOC devices to perform laboratory operations on a small scale are very appealing. First, the small volumes can reduce the time taken to synthesize and analyze a sample; the unique behavior of liquids at the microscale allows simpler control of molecular concentrations and interactions; and reagent costs and the amount of chemical waste can be much reduced. Compact devices also allow samples to be analyzed at the point of need rather than in centralized laboratory [273].

[1] Parts of the results have been published in Ref. [270, 271].

The simple fabrication of DFB-OSLs makes them very attractive for the integration in miniaturized spectroscopic systems. Especially, they can be fabricated on flexible polymer substrates and can operate as pixel-sized light sources in an all-polymer LOC platform. A good example is the LOC fluorescence marker detection scheme demonstrated by Vannahme et al. [30, 274], which comprised an integrated organic DFB laser, an optical waveguide obtained by index modification of poly(methyl methacrylate) (PMMA) under deep ultraviolet (DUV) illumination and the microfluidics with fluorescent molecules.

However, for such LOC platforms the size of DFB lasers should be as small as possible to save the limited space for other passive optical components. Additionally, there must be an accurate lateral control to define the localized laser sites. In order to realize such "laser pixels", techniques for local definition of the active medium and for laterally controlled fabrication of DFB-corrugations have to be developed.

In terms of the active media, small molecules such as Alq_3 doped with the laser dye DCM are competing with conjugated polymers. The former class of materials can be evaporated locally using shadow masks and can thus be combined with pre-defined grating sites [18]. On the other hand, conjugated polymers can be processed by inexpensive solution-based techniques such as spin coating, dip-coating and doctor blading. However, these methods result in an organic semiconductor layer covering the whole substrate. An appropriate solution processing technique to create conjugated polymer laser pixels is still in vacancy. To solve this problem, printing techniques can be used to achieve an accurately controlled deposition. A particularly interesting approach is the use of ink-jet printing.

In terms of the DFB-corrugations, the laterally controlled fabrication of grating structures for organic semiconductor lasers has only been addressed so far by using nanograting transfer to transfer the gratings onto a homogeneous gain material layer [275] and a rather time-consuming approach using direct electron beam lithography (EBL) on conjugated polymers [276]. On the other hand, various substrate scale techniques such as laser interference lithography (LIL) [277] and nanoimprint lithography (NIL) [243] are available to achieve a low-cost fabrication of high-quality DFB corrugation nanostructures on an LOC platform. As introduced in Section 3.1.2, to solve the problems in fabrication of pre-defined nanogratings by using state-of-the-art NIL techniques, a novel method, laser-assisted replication, can be adopted in the process as a substitute for hot embossing.

4.2 Organic semiconductor lasers fabricated by ink-jet printing

Ink-jet printing has already been successfully applied in the fabrication of various organic electronic devices, such as organic transistors [260, 278], organic solar cells [259, 279] and organic light emitting diodes [261, 280], but not yet for building organic semiconductor lasers before author's work. In contrast to most solution-based deposition techniques, ink-jet printing allows the deposition of thin films with arbitrary lateral shapes and small lateral dimensions [281]. Thus, ink-jet printing is a promising technique for the inexpensive fabrication of spatially defined organic lasers ("laser pixels") in photonic LOC or other sensing systems. As it allows printing several different organic materials onto one substrate, ink-jet printing is also capable to improve the tunable range of organic lasers in the visible spectrum for free-space applications.

In this section, ink-jet printing of organic semiconductor distributed feedback lasers from polymer solution is presented. An example of precise printing on several pre-defined grating sites in small lateral dimensions is given to verify the fabrication advantage. Subsequently, the uniformity of the printed organic layer and of the resulting laser emission is investigated.

4.2.1 Ink-jet printing process and materials

For printing the active layers an ink-jet printer (FUJIFILM Dimatix, Dimatix Materials Printer DMP-2800) with piezoelectric driven nozzles in the printing head (DMC-11610) was applied.[2] The patterns to be jetted were

[2]The ink-jet printer is located at InnovationLab GmbH in Heidelberg.

resolved into 5 µm pixels (5080 dpi) from silicon nozzles with an orifice size of approximate 21.5 µm. The jetting duration, frequency, slew rate and firing voltage were optimized in one synthetic jetting waveform according to the characteristics of the adopted solutions. In this work, only one nozzle from the linearly spaced 16 nozzles was used for fabrication to avoid complexity due to overlapping printing of neighboring nozzles and jetting discrepancies of different nozzles.

The printability of organic semiconductor lasers was investigated in respect of organic solvents and conjugated polymers. Solvents with low boiling points, like the frequently used toluene and tetrahydrofuran (THF), are not suitable for ink-jet printing due to the fast evaporation at the nozzle of cartridge, which impedes proper droplet formation. In this work, a solvent blend of one high-boiling solvent ethyl benzoate and one low-boiling solvent toluene was applied to avoid such clogging problem and to minimize the coffee stain effect [282]. Furthermore, the concentration of the dissolved conjugated polymer, which was frequently used over 20 mg ml^{-1} for organic laser deposition [22, 39], has to be decreased. As the solute concentration gets higher, the viscosity of the solution increases, while the surface tension decreases [283, 284]. Both effects are detrimental for a steady printing process. 10 mg ml^{-1} of the conjugated polymer $F8_{0.9}BT_{0.1}$ was dissolved in a solvent blend of ethyl benzoate and toluene (20:80 vol.%/vol.%) for the following printing process.

4.2.2 Ink-jet-printed organic laser pixels

As shown in Figure 4.1(a) and 4.1(b), individual active pixels were ink-jet-printed on several grating sites on a TOPAS® 6013 cyclic olefin copolymer (COC) (TOPAS Advanced Polymers) substrate with a thickness of 254 µm, which were obtained by replicating 70-nm-deep distributed feedback gratings with site dimensions of 0.5 × 0.5 mm^2 from a silicon stamp using hot embossing. The DFB-corrugations were designed as gratings with a period of 360 nm in 1 µm-deep micro-trenches. A thin layer of hafnium oxide (HfO_2) with a thickness of 80 nm was deposited on top of the COC substrates in order to protect the polymer substrate from erosion of the employed solvents. The prepared solution was ink-jet-printed with a drop spacing of 25 µm (1016 dpi) onto the grating sites. The fast axis of the print head movement was aligned parallel to the grating orientation (the x-direction as shown in Figure 4.1(a)). To accelerate the evaporation rate of

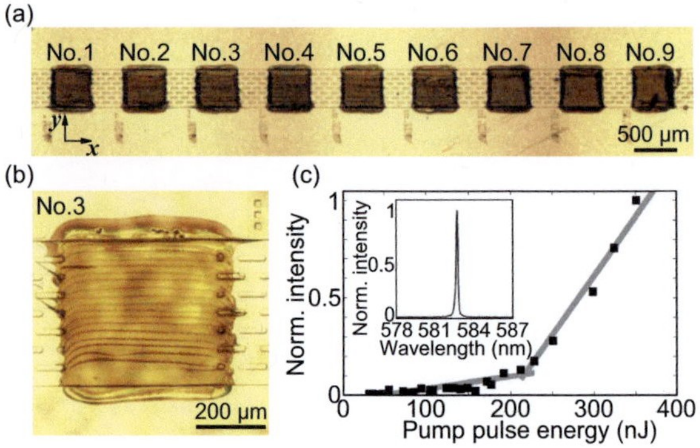

Figure 4.1: (a) Optical micrograph of ink-jet-printed organic lasers on 9 separate TOPAS® 6013 COC grating sites. (b) Individual micrograph of organic laser printed on grating site No.3. (c) Laser threshold and emission spectrum (inset) of the ink-jet- printed organic DFB laser on grating site No.3.

the solvent, the substrate was heated to 60 °C, while the nozzle was kept at room temperature. The mean thickness of single deposited organic films was 650 nm with a standard deviation of about 65 nm as measured with a surface profiler (Bruker, DektakXT^{TM}).

The optical characterizations were performed with the setup at LTI (see Section 3.3). The laser threshold was measured as 210 nJ pulse^{-1} with an elliptical pump spot with a diameter of 81 µm parallel and 46 µm perpendicular to the grating orientation. The input-output characteristic of ink-jet-printed laser is shown in Figure 4.1(c) with a laser spectrum in the inset. The polarization of laser emission at 582.7 nm was parallel to the grating orientation, which verified a surface-emitting laser of resonant TE-mode [285]. The whole printed grating area was scanned with a pump spot size of 81 µm × 46 µm and the laser spectra were recorded at each measurement point with a scanning spacing of 100 µm. The results revealed a mean laser wavelength of 582.3 nm and a standard deviation of about 0.4 nm, which indicated a small variation in the film thickness on the length scale of the laser pixels. At a pump energy of 400 nJ pulse^{-1} and a repetition rate of 1 kHz, the device operated continuously for about 3×10^6 pump pulses before decaying to 50 % of the initial laser intensity (50 min operation). This laser lifetime is comparable to one of the similar $F8_{0.9}BT_{0.1}$ DFB lasers fabricated by spin-coating or horizontal dipping.

4.2.3 Uniformity of ink-jet-printed organic lasers

To investigate the uniformity and the lasing behavior of ink-jet-printed organic DFB lasers, a relatively large area (2 mm × 3 mm) of $F8_{0.9}BT_{0.1}$ was printed with a drop spacing of 30 µm (847 dpi) on a 1D-grating substrate with a grating period of 350 nm. A uniform film was deposited by applying the described parameters on the same site for two printing cycles, as shown in Figure 4.2(a). The thickness measurement revealed an average thickness of 700 nm without noticeable coffee staining effect, as presented in Figure 4.2(b). Nevertheless, the strip-like printing profile can be seen due to the intrinsic line-by-line deposition operation, as shown in the atomic force microscope (AFM) image in Figure 4.2(c). A roughness analysis of the surface revealed that the averaged root-mean-square (RMS) roughness in the x-direction is about 8 nm, while it is about five times higher in the y-direction. This reveals that the printing uniformity along the fast axis of the print head movement (x-direction) is considerably higher than that along the perpendicular axis (y-direction).

Figure 4.3(a) shows TE-mode lasing spectra with a peak wavelength of 563.5 nm for two different pump spot sizes at two different excitation energy levels. At a pump spot size of 48 µm × 66 µm, the laser threshold was measured as 76 nJ pulse^{-1}. The spectral width of the laser emission increased with the enlarged pump spot sizes (S_P) and the enhanced pump pulse energy (E_P). The emission broadening along with the enlarged pump

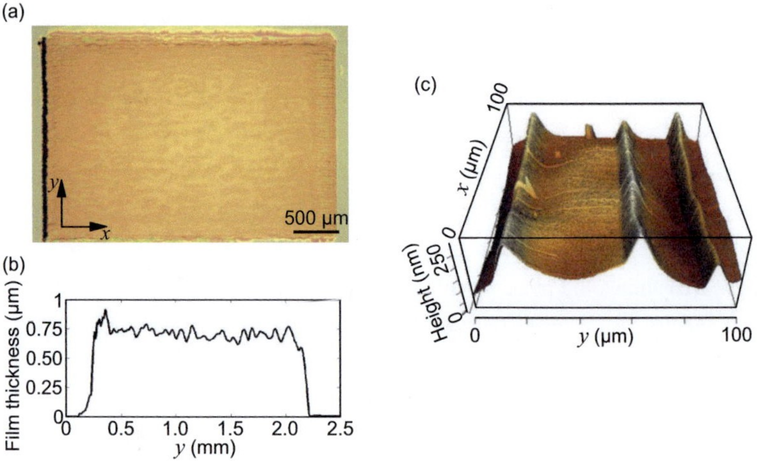

Figure 4.2: (a) Optical micrograph of a conjugated polymer film with a lateral dimension of 2 mm × 3 mm printed on a large-area SiO_2 grating. (b) Film thickness profile measured along the *y*-direction with a surface profiler. (c) Atomic force micrograph of the ink-jet-printed surface profile with an area of 0.1×0.1 mm^2.

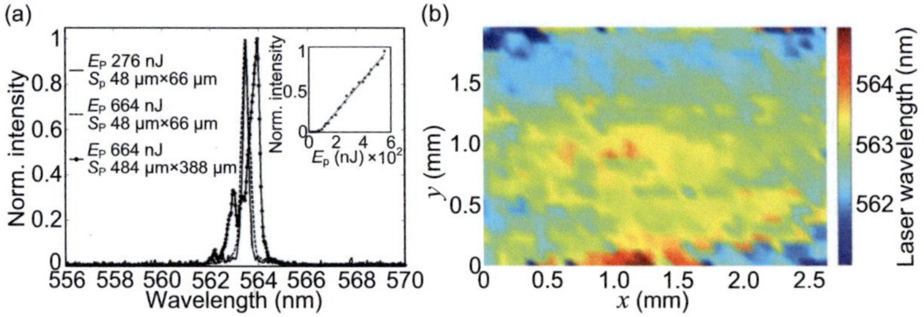

Figure 4.3: (a) TE-mode emission spectra of an ink-jet-printed organic laser under excitations with varying spot sizes and input-output characteristic (inset) for excitation with a pump spot size of 48 µm × 66 µm. E_P: pump pulse energy, S_P: pump spot size. (b) Color-encoded spatially resolved laser wavelengths for the TE-mode of an organic DFB laser, printed with a resolution of 30 µm (847 dpi).

spot sizes was mainly attributed to the inhomogeneity of film thickness within the excitation area and not to the increasing number of lateral modes of DFB lasers with linear gratings [116], since the spectral broadening was in both of longer-wavelength and lower-wavelength directions. The laser emission uniformity was checked by scanning the whole deposition area with a constant excitation intensity of 400 nJ pulse^{-1} and a pump spot size of 81 µm × 46 µm. The scanning spatial resolution was 75 µm. Figure 4.3(b) shows the peak wavelength distribution of the spatially resolved laser spectra. An average laser wavelength was measured as 562.9 nm with a standard deviation of 0.5 nm.

4.3 Organic semiconductor lasers fabricated by laser-assisted replication

As elucidated in Section 3.1.2, a spatial replication of a DFB-corrugation for fabrication of organic laser pixels can be achieved using a near-infrared high-power diode laser beam. In this section, the novel laser-assisted replication is demonstrated to fabricate localized surface-emitting (second-order) and edge-emitting (first-order) organic distributed DFB laser pixels on a poly(methyl methacrylate) (PMMA) substrate. This technique allows a fast replication with standard polymer materials. Not only the process time is reduced, the localized heating also provides the unique benefit to replicate parts of existing structures on a master stamp. Additionally, since only the defined areas are to be heated, it allows a subsequent additive fabrication of DFB-corrugations while all other areas are retained. The minimum localized replication area and the influence of processing parameters on corrugation quality are investigated in Section 4.3.4.

4.3.1 Fabrication process

The basic principle of laser-assisted replication was already introduced in Section 3.1.2. Before replication, a silicon master stamp with a DFB-corrugation was fabricated. Via electron beam lithography (EBL) and reactive ion etching (RIE), the DFB-corrugations with grating periods ranging from 195 nm to 450 nm, a height of ~ 120 nm and a duty cycle of 75 % were generated. Via aligned UV-lithography and subsequent RIE on the same wafer, trenches for the DFB lasers with a height of 1.2 µm were finished. A thin anti-adhesion layer was then deposited on the wafer to facilitate the polymer substrate detachment. For the replication process, a 2-mm-thick PMMA substrate (Evonik, Plexiglas® XT) with a size of 30×30 mm^2 was positioned on selected corrugations of the silicon master stamp. The following processing parameters resulted in highest quality replication: a heating temperature of $T = 180$ °C, a laser scan velocity of $v = 40$ mm s^{-1}, a clamping pressure of $p = 0.1$ MPa and a line offset (writing distance between two subsequent lines) of $\Delta x = 0.3$ mm. To achieve a homogenous replication, each substrate was treated twice by the laser beam: first perpendicular to grating orientation and second parallel to the grating orientation. The replication process took about 5 min for the replication of an area of 30×30 mm^2. Figure 4.4(a) and 4.4(b) show a microscope image and a scanning electron microscope (SEM) image on a replicated PMMA substrate with grating period of 400 nm.

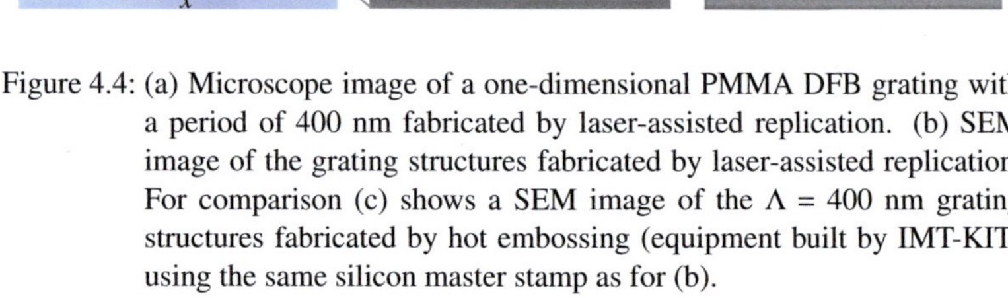

Figure 4.4: (a) Microscope image of a one-dimensional PMMA DFB grating with a period of 400 nm fabricated by laser-assisted replication. (b) SEM image of the grating structures fabricated by laser-assisted replication. For comparison (c) shows a SEM image of the $\Lambda = 400$ nm grating structures fabricated by hot embossing (equipment built by IMT-KIT) using the same silicon master stamp as for (b).

The grating structures fabricated via laser-assisted replication shows a high fidelity over the whole molding area. The replicated quality was similar to hot embossing, as shown in Figure 4.4(c). However, the laser-assisted replication process took a much shorter time compared to the 45-min-long hot embossing process.

After replication, the clamping pressure was released. The PMMA substrate was cooled down in ambient condition and separated from the silicon master stamp. Alq_3 doped with the laser dye DCM has been chosen as the active material. As elucidated in Section 2.2.4, this small molecule blend forms a very efficient Förster energy transfer system and exhibits an excellent long-term stability. Evaporating 250 nm Alq_3 doped with 2.8 wt.% DCM onto the defined grating sites finally yielded the organic DFB laser pixels. The used stencil shadow mask was fabricated by UV-lithography and nickel electroplating [30], as described in Section 3.2.1.

4.3.2 Second-order organic DFB laser pixels

The optical characterization setup is the same as that for the ink-jet-printed organic laser pixels (see Section 3.3). Figure 4.5(a) shows the color-encoded spatially resolved lasing wavelength from a PMMA chip containing 9 laser pixels with individual lateral dimension of $0.5 \times 0.5\ mm^2$ and an interspacing of 0.5 mm. The grating period of DFB-corrugations varies from 370 nm to 450 nm in steps of 10 nm from the leftmost pixel to the rightmost one. The chip was probed with a resolution of $100\,\mu m$. The laser emission wavelength did not show any deviation on the area of a single laser pixel within the spectral resolution of the setup (approximately 0.1 nm). This indicates a very uniform replication of nanogratings in PMMA. The corresponding laser emission spectra and the threshold fluence values for TE_0-modes are shown in Figure 4.5(b) and 4.5(c). After confirmation using a polarization filter, the TM_0-mode laser peaks can also be detected at $\lambda = 630.4$ nm and $\lambda = 643.6$ nm corresponding to gratings of $\Lambda = 400$ nm and $\Lambda = 410$ nm. The higher laser threshold from the laser pixel with a grating period of 450 nm is attributed to the lower gain coefficient at the emission wavelength of 712.5 nm.

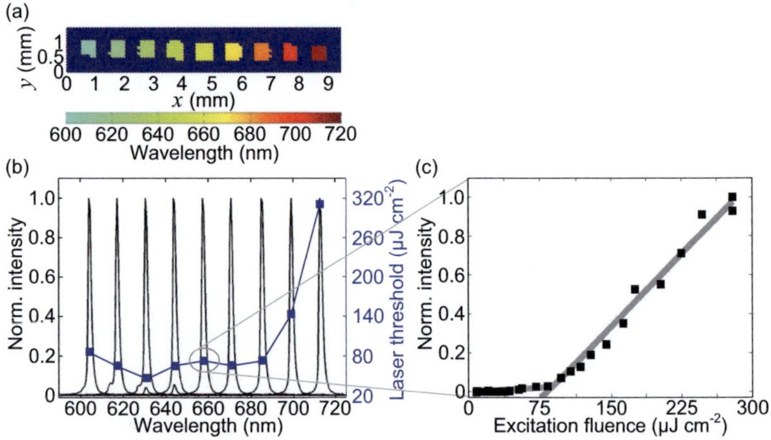

Figure 4.5: (a) Color-encoded spatially resolved laser wavelengths of fabricated organic DFB laser pixels with grating periods from 370 nm (left) to 450 nm (right) in steps of 10 nm. (b) Laser spectra and TE_0-mode threshold fluence measured at a pump spot size of 220 µm × 170 µm corresponding to the grating distribution. For some gratings the onset of the TM_0-mode can also be seen as very small peaks. (c) Input-output characteristic of a laser pixel with grating period of 410 nm and a laser emission at λ = 657.5 nm.

4.3.3 First-order organic DFB laser pixels

For being suitable for LOC spectroscopic applications, the organic semiconductor lasers are usually based on first-order DFB gratings for emission light in chip-plane only. By introducing a waveguide onto the chip, first-order organic laser emission can be efficiently coupled and guided to the analyte sites. PMMA chips with deep ultraviolet (DUV) induced waveguides have been fabricated previously [18, 274]. Here, the active grating and waveguides were fabricated altogether via laser-assisted replication and Alq_3:DCM was evaporated on both of them to build coupled edge-emitting organic laser on chip. This novel configuration allows a one-step fabrication of laser and waveguide without mix-and-match pattern-related defects [286]. Additionally, direct connection between laser and waveguide made from the same active material facilitates good coupling efficiency.

Different gratings with length of 0.5 mm, interconnected by 0.3-mm-wide waveguide structures perpendicular to the grating orientation were simultaneously fabricated in 1.2 µm-deep trenches on PMMA chip. The grating periods for first-order organic DFB lasers were chosen to be 195 nm, 200 nm, and 205 nm, as labeled in Figure 4.6(a). A 250-nm-thick layer of Alq_3:DCM was evaporated on the defined grating and waveguide, thus allowing an optimum light coupling.

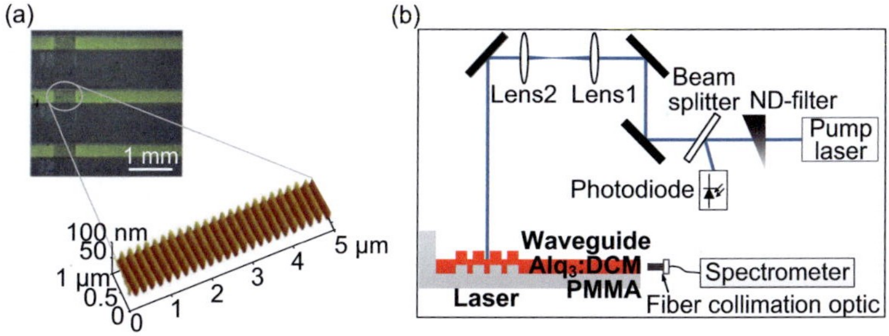

Figure 4.6: (a) Microscope image of a coupled organic DFB edge-emitting laser and an AFM image of the nanograting with period of 200 nm. (b) Schematic illustration of the optical setup to detect the edge emission from the integrated organic DFB laser.

The experimental setup for the chip characterization is schematically depicted in Figure 4.6(b). The pump laser and the detection system were the same ones as used for the characterization of the surface-emitting organic lasers, but the detection plane was changed according to the chip orientation. The excitation spot diameter on the sample surface was $\sim$ 100 µm. In order to limit laser degradation in the ambient atmosphere, the excitation repetition rate was set as low as 100 Hz.

Figure 4.7(a) shows exemplarily the laser spectrum of a DFB laser with a grating period of 200 nm. The laser emission at λ = 645.5 nm was detected at the chip facet using a multi-mode optical fiber (Ocean Optics, P400-3-UV-VIS). A stronger background due to the Alq_3:DCM fluores-

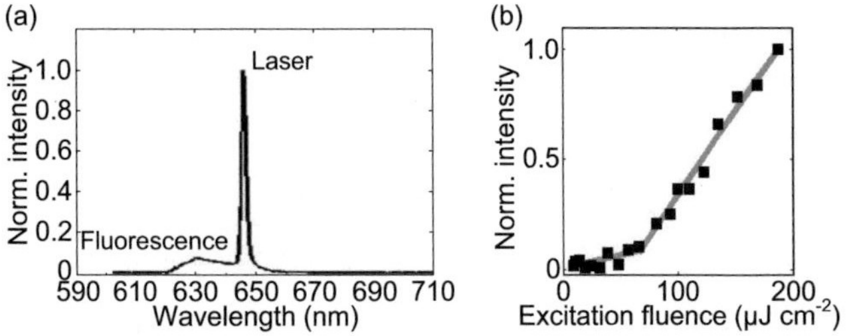

Figure 4.7: (a) Laser spectrum and (b) input-output curve taken from the chip facet for a grating with $\Lambda = 200$ nm using a multi-mode optical fiber coupled to a spectrometer.

cence than in the case of the surface-emitting lasers was observed. This can be explained by the fact that the fluorescence was coupled into the waveguide with comparable efficiency as the laser radiation. In case of surface-emitting lasers, the fluorescence signal was weaker as it emitted into all directions and only the perpendicularly emitted light was detected. The corresponding laser threshold fluence was measured to be 72 µJ cm^{-2} (see Figure 4.7(b)).

In comparison with the previously used DUV-induced PMMA waveguides, the on-chip waveguides from the same organic laser material are in greatly advantageous in terms of coupling efficiency. However, the asymmetrical waveguide scheme (from top to down: air, Alq_3:DCM and PMMA) may influence the propagating laser mode profile.[3] It is also noted that the waveguides should be designed as short as possible to reduce the self-absorption of organic active media in the waveguides.

4.3.4 Size-dependency of laser-assisted replication

The minimum replication area size of laser-assisted replication is determined by the applied laser beam diameter, the processing time, and the laser-induced heating temperature in the contact zone of master stamp and

[3]The DUV-induced PMMA waveguides can be considered as a symmetrical waveguide after chip bonding with a PMMA lid [30].

polymer. To investigate the spatial resolution limits of the replication process, test structures were fabricated with a tool comprising spatially defined DFB grating areas using different heating periods. Figure 4.8(a) shows a microscope image of a beginning trail of the replicated structure, which was obtained by laser-induced heating of a single line with a scanning speed of 40 mm s^{-1}, corresponding to a local heating as short as 17.5 ms. The red line indicates a fit to the right end of the replicated structures.

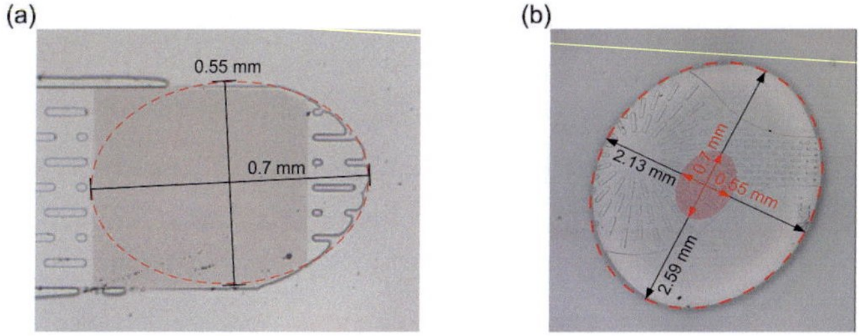

Figure 4.8: Microscope images of (a) the begin of the replication trail obtained by heating a single line with a laser scanning velocity of 40 mm s^{-1} and (b) the replicated spot after 2-s-long punctual heating in comparison with the laser focus spot (red).

The rim is in good accordance with the elliptical laser focus spot size of 0.55 mm $\times$ 0.7 mm. In comparison, applying 2-s-long punctual heating resulted in an expansion of the heat-affected zone on the silicon stamp and a consequently enlarged replication area. As shown in Figure 4.8(b), an extension was measured as $\sim$ 3.8 times of the laser focus spot diameter.

The versatility of the localized replication was verified by fabrication of pre-selected laser pixels. Figure 4.9(a) shows a partial replication from the silicon master stamp with gratings in periods of 390 nm, 400 nm, 410 nm and 420 nm. Each pixel has a dimension of 0.5 $\times$ 0.5 mm^2 with an interspacing of 0.5 mm. A 2-mm-broad space was excluded from the laser writing path from left to right on the chip. As a result, an unstructured area with a width of 1.8 mm was observed between the replicated structures. The slight difference between the size of unstructured area and the

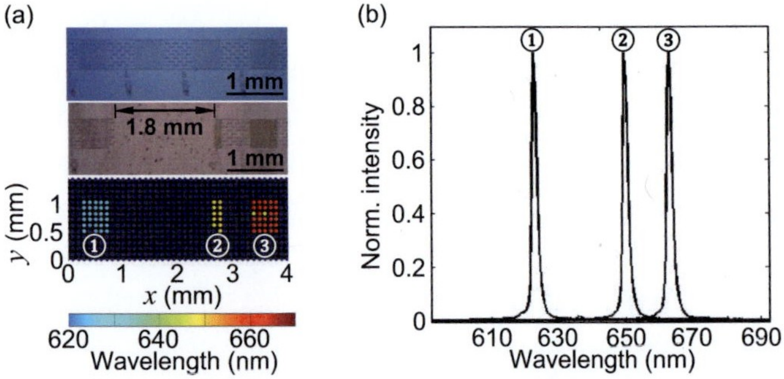

Figure 4.9: (a) Photograph of the complete set of spatially defined laser gratings (from left to right: $\Lambda = 390$ nm, $\Lambda = 400$ nm, $\Lambda = 410$ nm and $\Lambda = 420$ nm) on silicon master stamp (top). The actual laser grating areas are visible as homogenous darker grey areas which are separated by auxiliary structures in between. The photograph of the PMMA laser pixels partially replicated on a PMMA chip (middle) and spatially resolved laser wavelengths of fabricated organic DFB laser pixels (bottom) are also shown. (b) Laser spectra corresponding to the grating areas ①, ② and ③ in (a).

non-irradiated area is attributed to the heat conduction in the silicon master stamp. The spatially resolved lasing wavelength distribution revealed one completely omitted and one only partially available laser pixel. The latter had a clearly defined border. The corresponding laser pixel spectra are shown in Figure 4.9(b).

As previously mentioned, a great advantage of localized replication is the subsequent fabrication of nanostructures with already existing photonic, fluidic or electronic components, which remain unaffected during fabrication process. This was demonstrated by replicating the same structures twice on a single 2-mm-thick PMMA substrate and performing two successive replication processes at a distance of 10 mm. Each replicated area was defined to be 5 mm $\times$ 20 mm without overlapping each other, as shown in Figure 4.10(a). Figure 4.10(b) shows a photograph of the finished laser pixels on PMMA chip, which can be clearly identified by the grating effect. A 250-nm-thick Alq_3:DCM layer was evaporated on top of the chip to investigate the laser behavior and to determine the replication quality. All laser pixels showed homogeneous laser emissions throughout the whole pixel area. Laser pixels having the same grating period revealed comparable characteristics, e.g., the threshold fluences of laser pixels with a grating period of 420 nm were measured to be 61 µJ cm^{-2} and 52 µJ cm^{-2}, respectively. Both laser pixels emitted at the same laser peak wavelength of 670 nm. The existing grating structures were not influenced by the subsequent replication process. Inexpensive fabrication of identical nanostructures in a step-and-repeat process is shown to be possible. This can be widely useful for LOC and other integrated optics applications.

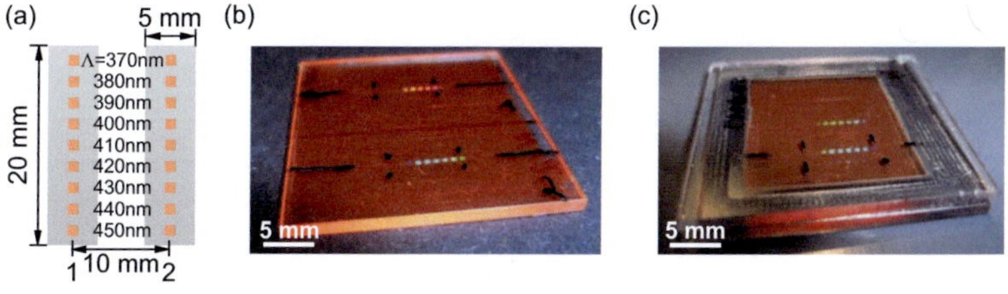

Figure 4.10: (a) Scheme of the first and second laser-assisted replication area in size of 5 mm × 20 mm. (b) Photograph of subsequently fabricated grating structures 1 and 2. (c) Finished chip after encapsulation with PMMA lid via laser transmission welding. The chip edge looks silver due to the carbon absorption layer.

4.3.5 Encapsulation via laser transmission welding

During operation, organic semiconductors suffer from photo-induced degradation when exposed to air or water. Therefore, an encapsulation is necessary to prolong the device lifetime. Earlier, LOCs containing organic semiconductor lasers were encapsulated in polymer using thermal bonding [274, 287], which is rather time-consuming and therefore only suitable for parallel processing with large numbers of LOC chips. In addition, the high temperature used for bonding leads to bleaching of the organic active media. Here, the encapsulation of the fabricated organic laser chip is achieved via laser transmission welding [288]. For this purpose, the same setup as for laser-assisted replication was used. A 2-mm-thick PMMA substrate was prepared as a lid and placed on top of the previously fabricated chip with laser-pixel trenches. Since PMMA is transparent for near-infrared radiation, an additional absorption layer of carbon with a thickness of 5–10 nm was sputtered at the square-edge on the surface of the PMMA lid pointing towards the laser chip, where the welding process was to be performed. Applying the diode laser beam at a wavelength of 940 nm and a maximal power of 50 W on the chip edge with 5-mm-wide margin, the absorption layer converted the laser radiation into heat, which melted the two PMMA substrates along the contact area and bonded them together (T = 140 °C; v = 40 mm s^{-1}; p = 0.4 MPa; Δx = 0.3 mm). A photograph of the encapsulated organic laser chip is shown in Figure 4.10(c). As expected, the encapsulation process was found not to affect the emission properties of the laser pixels. Laser transmission welding is thus not only a practical tool for chip bonding directly after laser-assisted replication, but it also can substitute thermal bonding after a hot embossing process.

4.4 Conclusion

In this chapter, novel fabrication approaches for spatially defined organic semiconductor lasers were presented. The laser devices can be potentially applied as integrated pixel-sized light sources into a lab-on-a-chip (LOC) platform. Two technologies, ink-jet printing for the local definition of active materials and laser-assisted replication for the laterally controlled fabrication of DFB-corrugations were developed.

To employ ink-jet printing in fabrication of organic semiconductor lasers, a mixture of high-boiling and low-boiling solvents was utilized for dissolving the polymer to optimize the film uniformity. In contrast to most solution-based deposition techniques, ink-jet printing allows the deposition of thin films with arbitrary lateral shapes and small lateral dimensions. Thus, an accurate deposition of the organic gain medium on defined grating trenches without contamination of other parts of the functional chip could be achieved. The ink-jet-printed laser devices showed laser thresholds as low as 76 nJ pulse^{-1} and a spectrally homogeneous laser emission over large areas. The laser emission properties are comparable to those of organic semiconductor lasers fabricated via more established solution-processing methods.

The spatially defined organic laser pixels can also be realized by large-area deposition on pre-determined grating sites, which can be achieved by laser-assisted replication. As an alternative to the hot embossing technique, laser-assisted replication employs the irradiation of an infrared laser beam instead of large-area heating, thus allowing a shorter heating/cooling period and a localized pixel-sized replication. Based on this technique, surface-emitting laser pixels were realized based on nanogratings with different periods in lateral dimension of 0.5×0.5 mm^2 on a PMMA substrate. A simultaneous fabrication of nanoscale DFB-corrugations and neighboring microscale waveguide trenches was also demonstrated. In this way, edge-emitting organic lasers coupled to functional waveguides were realized. The minimum fabrication area size was investigated by comparing the replication spot with the laser beam focus size. The versatility of spatially localized replication was verified by fabrication of pre-selected laser pixels. It was also proven that using this fabrication method, further nanostructures could be added into the chip platform without negative influence on neighboring photonic components. Finally, the encapsulation of the polymer chip was shown using laser transmission welding against ambient air and water. It can be further improved with the help of a vacuum pump to achieve a complete isolation from oxygen.

The spatially defined grating sites fabricated by laser-assisted replication can be combined with the pixel-sized deposition of organic semiconductors via ink-jet printing. It is foreseen that by using various organic active media, this approach allows the digitally manufactured versatile coherent light sources integrated in photonic microsystems and LOC devices for spectroscopic applications.

5 Organic Semiconductor Lasers for Raman Scattering Measurements

Due to the narrow laser line width, and the low laser threshold, organic semiconductor distributed feedback lasers are good candidates for excitation sources for Raman spectroscopy. Some aspects of their laser characteristics need improvement to fulfill the requirements of Raman spectroscopy. This chapter presents firstly how to enhance the laser output power and maintain the stability of DFB lasers. In the following subsections, applications of organic semiconductor lasers in Raman spectroscopy are described. Two possible free-space excitation configurations, free-beam and fiber coupling, are investigated. To the best of the author's knowledge, this is the first application of organic semiconductor lasers in the field of Raman spectroscopy. In the final discussion the measurement results are compared and the possible ways for further improvement of this approach are given.[1]

5.1 Issues on organic laser properties

For Raman spectroscopy, the compact hybrid organic-inorganic laser diode could be a good substitute for bulky liquid dye lasers. Additionally, a very broad laser wavelength tunability can be obtained via fabricating an organic semiconductor laser by various means. A selection of an arbitrary excitation wavelength in the visible spectral range is of significant interest for resonance Raman scattering (RRS) and surface-enhanced Raman spectroscopy (SERS) [2, 290].

[1]Parts of the results have been published in Ref. [179, 275, 289].

As introduced in Section 2.4.2, adequate laser power, high wavelength stability, high spatial mode quality and moderately narrow line width are basic criteria for a laser for Raman spectroscopy. In an application-oriented view, the following issues ought to be under consideration to improve the performance of organic lasers for Raman measurements.

5.1.1 Laser threshold

The laser threshold is defined as the value of the pump energy at which the optical gain is beginning to surpass the losses of the laser resonator, as shown in Figure 5.1(a). A low laser threshold is desired because it can lower the pump energy and potentially increase the laser output power for a given pump energy.

DFB lasers are intrinsically low-threshold lasers. Typical threshold fluence values for organic second-order DFB lasers are on the order of 0.1–100 μJ cm^{-2} [68, 69]. A low threshold power requires basically low resonator losses and a high gain efficiency. The former is related to the DFB resonator parameters, such as grating depth, grating duty cycle, substrate refractive index (waveguide confinement) and so on. There are also some novel ideas, e.g., employing mixed-order grating structures or two-dimensional grating structures to lower the threshold value [166, 173, 174, 188]. The latter can be achieved by employing proper active media with a high overall optical gain. Furthermore, the excitation conditions, especially the pump spot size, determine the threshold values as well [291]. This will be discussed in detail in Section 5.2.3.

5.1.2 Laser output power

The optimization of laser output power for a given pump power and a given repetition rate usually involves a high slope efficiency η (see Figure 5.1(a)) and a low laser threshold E_{th}. Upon a high pump energy level, the slope efficiency is dominant in determining the laser output power. Additionally, pump pulse duration, pump repetition rate and stability of active media in-

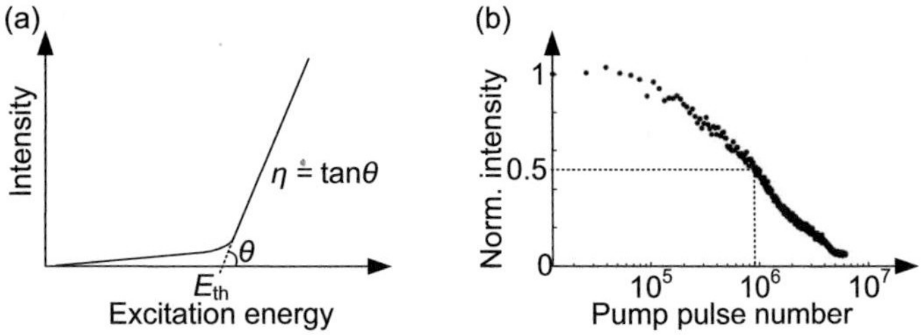

Figure 5.1: Characteristics of a laser: (a) The input-output characteristic shows a sharp bend at the threshold excitation energy E_{th} and the slope efficiency η. (b) An example of laser intensity degradation of an $F8_{0.9}BT_{0.1}$ DFB laser. The laser lifetime is here defined as the time period until the laser intensity degrades to half of its initial value.

fluence the laser output power as well. Due to the small excitation volume, the output pulse energy of organic semiconductor lasers is typically on the order of nanojoules, which corresponds to a peak power at the watt-level in a very short pulse duration (few nanoseconds). A very high organic laser slope efficiency of 70 % was reported in Fabry-Pérot waveguide laser using Alq_3:DCM as an active medium [15]. The differential slope efficiency of organic semiconductor second-order DFB lasers is in the range of 0.1–17 % [166, 186, 292, 293, 294, 295, 296]. The main approaches to enhance the slope efficiency are varying grating configuration heights [297, 298], using a two-dimensional DFB configuration [166, 169, 186], introducing an intermediate high refractive index polymer layer [299], and employing efficient organic active media [292, 296].

5.1.3 Laser lifetime

Like for all organic optoelectronic devices, photodegradation is a serious issue for organic semiconductor lasers and the operating stability of the devices is thus still a challenge [12]. Despite a number of studies on this subject [39, 93, 104, 300, 301, 302, 303, 304, 305], there is still no coherent understanding until now. There have been various criteria to scale and compare the degradation of organic lasers. The most widely used lifetime definition is the time period until the laser peak intensity degrades to half of its initial value [188, 306], as illustrated in Figure 5.1(b). This laser lifetime is determined by many parameters: the intrinsic chemical and optical properties of the used organic materials, the stability of the resonator features, the excitation density and pulse duration as well as the ambient conditions during fabrication and characterization. Therefore, the reported lifetime values can be hardly compared as almost every single experiment parameter plays a crucial role on the final results. Some common views for a better lifetime performance comprise isolating from ambient oxygen during characterization, applying short excitation pulses and low repetition rates, and choosing stable organic active media with less loss in triplet states. Among them, an encapsulation protecting the organic active media against water and oxygen has often been adopted in the device fabrication process [109, 179, 274, 306, 307]. With the help of an encapsulation based on optical adhesives or thermal bonding of a polymer lid, the organic laser lifetime is comparable to the lifetimes measured for a bare sample in a vacuum chamber. However, even in vacuum or nitrogen-filled environment organic lasers tend to degrade [188, 308].

5.1.4 Laser beam divergence

As described in Section 2.3.2, laser beam profiles are strongly dependent on the resonator geometry. For DFB-OSLs, the divergence is also related to the excitation power. The dimension of the laser divergence is not so important for certain applications like, e.g., sensing, but it is a major concern for lasers destined to be used as free-space excitation sources for spectroscopic applications [69]. Unfortunately, organic DFB lasers have so far only shown diffraction-limitation when a two-dimensional corrugation and a low excitation power (slightly above the laser threshold value) are applied [53]. The multiple lateral modes which are typically observed may lead to coupling deficiencies in optical fiber or power loss in passing through optical lenses or filters. For the application in spectroscopy, there may be unpredicted consequences on the spectroscopic characterization due to the enlarged laser spot size projected on the analyte samples.

5.2 Optimization of organic semiconductor lasers

Basically, a DFB-OSL is made up from three essential parts: optical pump source, DFB-corrugation resonator and organic active medium. Below the influence of each factor is discussed individually.[2]

5.2.1 Optimum laser active media

As discussed in Section 2.1, two categories of organic active media, conjugated polymers and small molecules can be used for organic semiconductor lasers. In this work, three often-used organic active media are compared to each other: the polyfluorene-based conjugated polymer $F8_{0.9}BT_{0.1}$, the composite polymer blend $F8_{0.9}BT_{0.1}$:MEH-PPV and the composite molecular system Alq_3:DCM. The former two material systems were processed with the solution-based method horizontal dipping and Alq_3:DCM was deposited through thermal evaporation. For comparison, laser devices with three active media were build on SiO_2 gratings[3] with a grating height of 90 nm. A grating with $\Lambda = 360$ nm was chosen for $F8_{0.9}BT_{0.1}$ laser and another with $\Lambda = 400$ nm for $F8_{0.9}BT_{0.1}$:MEH-PPV and Alq_3:DCM lasers.

Laser emission wavelength and line width

The spectral emission ranges of three laser devices were introduced in Section 2.2.4. Under an optical pumping at $\lambda = 355$ nm, an organic DFB laser with an arbitrary wavelength in the spectral range from 550 nm to 700 nm could be fabricated.

[2]The optical characterizations of DFB-OSLs were performed at LTI. The characterization setup was shown in Figure 3.6 in Section 3.3 and the spectral resolution of the employed optical characterization system is about 0.1 nm.

[3]The grating were fabricated using electron beam lithography, followed by reactive-ion etching on a thermally oxidized silicon wafer (see Section 3.1.1).

To match commercially available optical filters designed for use with the standard Nd:YAG laser or the helium-neon (He-Ne) laser, wavelengths of λ = 532 nm and λ = 632.8 nm are preferred for Raman experiments. Additionally for surface-enhanced Raman scattering (SERS) measurements, a laser wavelength over 600 nm is advantageous considering the absorption properties of metallic Au/Ag nanoparticles/thin layer. Unfortunately it is very hard to obtain an emission wavelength above ~ 700 nm using organic semiconductors due to severe triplet-triplet absorption [69].

The laser spectra of devices built with 220-nm-thick organic layers of $F8_{0.9}BT_{0.1}$, $F8_{0.9}BT_{0.1}$:MEH-PPV and Alq_3:DCM are depicted in Figure 5.2(a)–(c), which were acquired upon an optical pump energy of two times the corresponding laser thresholds. The laser spectral line width, often referred to as the full width at half maximum (FWHM) of the laser spectrum, is a comprehensive parameter which is dependent on the stimulated emission cross section of active medium, the laser emission wavelength as well as the pumping and resonator conditions [64]. The FWHM values of laser spectra in Figure 5.2(a)–(c) are listed in Table 5.1.

Properties \ Active media	$F8_{0.9}BT_{0.1}$	$F8_{0.9}BT_{0.1}$:MEH-PPV	Alq_3:DCM
Laser peak wavelength (nm)	552.6	604.5	633.1
FWHM (nm)	0.3	0.5	0.2
Threshold fluence (μJ cm^{-2})	58	20	4
Laser lifetime (10^6 pulses)	0.65	0.63	19
Wavelength shift $\Delta\lambda$ (nm)	0	−0.3	0
ΔFWHM (nm)	0	+0.2	0

Table 5.1: Laser characteristics of devices based on 220-nm-thick organic active layers of $F8_{0.9}BT_{0.1}$, $F8_{0.9}BT_{0.1}$:MEH-PPV and Alq_3:DCM. Pump spot size was set as 550 μm × 450 μm (2 × 10^{-3} cm^2). $\Delta\lambda$ and ΔFWHM represent the shift of laser peak wavelengths and the change in FWHM values during the laser lifetime measurements, respectively.

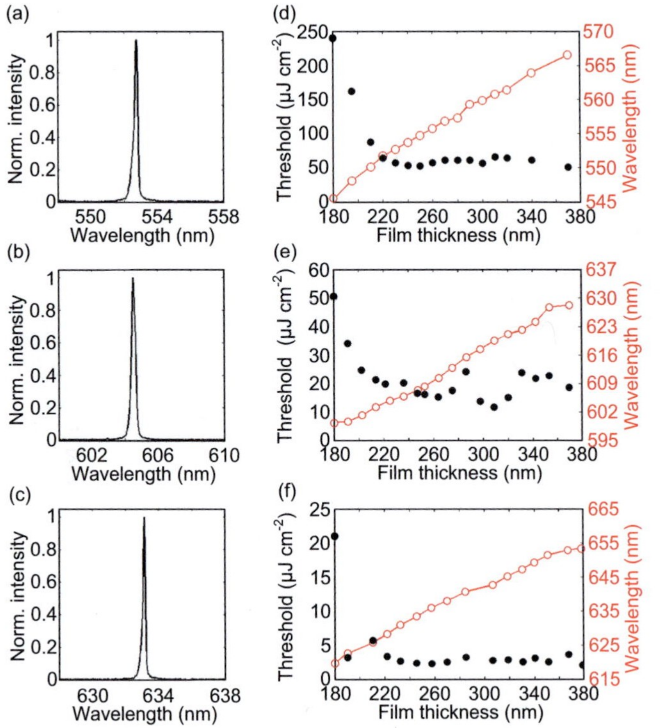

Figure 5.2: Laser spectra of organic laser devices with 220-nm-thick active layers of (a) $F8_{0.9}BT_{0.1}$, (b) $F8_{0.9}BT_{0.1}$:MEH-PPV and (c) Alq_3:DCM. The laser peak wavelengths are also listed in Table 5.1. The threshold fluence and laser emission wavelength as functions of organic film thickness for lasers based on (d) $F8_{0.9}BT_{0.1}$, (e) $F8_{0.9}BT_{0.1}$:MEH-PPV and (f) Alq_3:DCM.

Laser threshold and slope efficiency

For Raman spectroscopy, the laser device with lower laser threshold will potentially give a higher output power upon the same pump energy. By varying the active medium thickness, the laser threshold changes as well. The laser threshold fluences were measured for TE_0-like modes of laser devices with different active film thicknesses, but with the same pump spot size of 550 μm × 450 μm (2×10^{-3} cm^2), as shown in Figure 5.2(d)–(f). The results indicated that except for the initial points measured at film thickness smaller than ~ 200 nm, the laser thresholds stayed almost constant for the increasing thicknesses. For the thinnest film thicknesses, the waveguide mode has little overlap with the gain region of the organic materials. Less gain led to the higher laser thresholds. Taking advantage of the Förster energy transfer system, the composite polymer blend $F8_{0.9}BT_{0.1}$:MEH-PPV delivered a lower laser threshold than the sole conjugated polymer $F8_{0.9}BT_{0.1}$. Figure 5.3(a) shows the laser thresholds and the normalized slope efficiency of three laser devices. Deposited with the active films with identical thickness of 220 nm, the Alq_3:DCM laser device delivered the highest laser slope efficiency of ~ 5 % and the lowest laser threshold of 4 μJ cm^{-2}.

In general, utilizing the similar DFB grating substrates, the laser devices based on the composite small molecule system Alq_3:DCM demonstrated a lower laser threshold and a higher laser slope efficiency than devices based on $F8_{0.9}BT_{0.1}$ and $F8_{0.9}BT_{0.1}$:MEH-PPV.

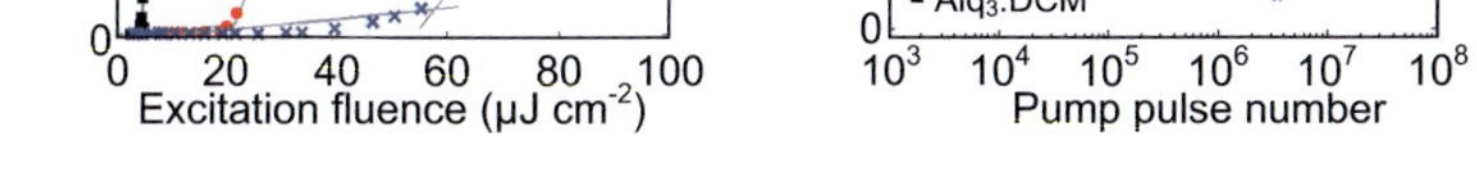

Figure 5.3: (a) Input-output characteristics of the organic DFB lasers built with 220-nm-thickness organic active media at laser wavelength of 552.6 nm ($F8_{0.9}BT_{0.1}$), 604.5 nm ($F8_{0.9}BT_{0.1}$:MEH-PPV) and 633.1 nm (Alq_3:DCM). The laser threshold fluence values are listed in Table 5.1. (b) Degradation characteristics of DFB lasers based on the three organic active media.

Laser emission stability

The photostability of DFB-OSL is in general a serious issue and especially for Raman measurements which may require long acquisition times. The photobleaching phenomenon is mainly caused by chemical modification of organic molecules, which leads to an irreversible loss in quantum efficiency and gain. The degradation is mainly evident in the decay of the laser peak intensity, the shift of laser emission wavelength and the broadening of the laser spectrum.

Firstly, the decays of the laser peak intensity of devices based on $F8_{0.9}BT_{0.1}$, $F8_{0.9}BT_{0.1}$:MEH-PPV and Alq_3:DCM were followed by monitoring the output power while maintaining the pump pulse energy constant. For the experiments the lasers were kept in a vacuum chamber (see Section 3.3). To exclude other influencing factors, laser lifetimes were measured with an identical optical pump spot size of $550\,\mu m \times 450\,\mu m$ (2×10^{-3} cm^2) and an optical pump energy of two times the corresponding laser threshold. The degradation characteristics of three laser devices are shown in Figure 5.3(b). The laser lifetime values are also listed in Table 5.1. Here the laser lifetime is defined as the elapsed number of pump pulses before the laser emission intensity drops to 50 % of the initial value. The composite small molecule system of Alq_3:DCM exhibited the best performance in laser power stability. Its lifetime was longer than $\sim 10^7$ pulses and exceeded the lifetimes of the polymer lasers by more than an order of magnitude.

The peak wavelength shifts ($\Delta\lambda$) and the FWHM changes (ΔFWHM) of laser spectra are also presented in Table 5.1. The two values for comparison were taken as the initial and the final ones in the corresponding laser lifetime measurements.

As a short summary here, the composite small molecule system Alq_3: DCM demonstrates an outstanding behavior in laser threshold, laser line width and laser emission stability. In addition, through deposition of a defined film thickness, the Alq_3:DCM laser device can provide a laser excitation wavelength exactly at 632.8 nm to match the commercial available optical filters designed for the He-Ne laser. Thus, Alq_3:DCM is a preferred laser active medium to build organic lasers for Raman experiments.

5.2.2 Optimum DFB-corrugation substrates

Substrate materials

A high contrast in refractive index between substrate and active medium is preferred for a laser device with low laser threshold and a high laser slope efficiency. Thus a DFB-corrugation substrate with a low refractive index is preferred. Glass gratings were first applied in the fabrication of organic semiconductor DFB lasers with advantages of easy cleaning process and high stability. Along with the advent of nanoimprint and laser interference lithography, the polymer DFB substrates, e.g., poly(methyl methacrylate) (PMMA) and cyclic olefin copolymer (COC), were becoming favorable materials for low-cost fabricated laser devices due to their high flexibility.

Substrate materials / Properties	SiO_2	Glass	PMMA	COC	ARN	SU-8
Substrate thickness (µm)	1[4]	1100	2000	1000	1.6[5]	5[6]
Refractive index (@ 633 nm)	1.46	1.52	1.49	1.53	1.49	1.59
Transmission (@ 355 nm)	92%	>90%	81%	82%	<10%	20%
Transmission (@ 633 nm)	92%	>90%	92%	91%	>90%	98%
Grating depth (nm)	90	90	120	120	140	20
Duty Cycle	50%	75%	80%	80%	50%	50%
Laser peak wavelength (nm)	633.1	631.7	628.7	632.4	628	642.9
FWHM (nm)	0.15	0.2	0.3	0.35	0.3	0.6
Threshold fluence (µJ cm^{-2})	4	15	8	8	32	38
Laser lifetime (10^6 pulses)	19	97	0.87	3.1	4.4	6.4

Table 5.2: Optical properties, grating features and laser characteristics of devices built on different DFB-corrugation substrates.

[4] ~ 1 µm oxidized layer on a 400 µm-thick silicon wafer substrate.
[5] Thickness of the photoresist SX AR-N 4800/16 layer on a 1-mm-thick glass substrate.
[6] Thickness of the photoresist SU-8 2005 layer on a 1-mm-thick glass substrate.

To compare the laser emission quality on different grating substrates, 220-nm-thick Alq_3:DCM thin layers were evaporated on various one-dimensional second-order grating substrates with a grating period of 400 nm. They were opaque SiO_2 surface gratings on a thermally oxidized silicon wafer fabricated via electron beam lithography (see Section 3.1.1), a commercial available D 263® glass grating substrate (Visolas GmbH), PMMA (Evonik, Plexiglas® XT) and COC (Topas 6013) grating substrates replicated by hot embossing using the same silicon template and grating substrates fabricated using laser interference lithography directly on photoresist materials of SX AR-N 4800/16 (Allresist) and SU-8 2005 (MicroChem) (see Section 3.1.3 and Appendix A.2). The DFB-corrugation properties of all substrates are listed in Table 5.2 [309, 310, 311, 312, 313].

Figure 5.4 presents the AFM micrographs of the surface corrugations of these grating substrates.

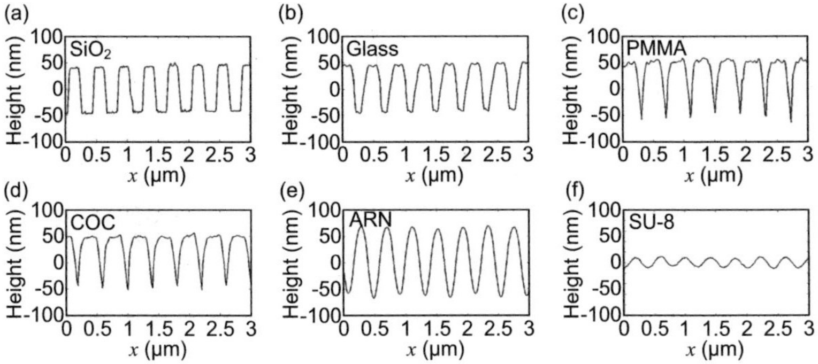

Figure 5.4: AFM micrographs of the surface corrugations on (a) SiO_2, (b) glass, (c) PMMA, (d) COC, (e) SX AR-N 4800/16 and (f) SU-8 grating substrates.

Owing to the difference in refractive index, the laser peak wavelengths of various grating substrates deviate from each other. Upon an optical pump energy as high as two times of the corresponding laser threshold, a minimum FWHM value of 0.15 nm was achieved from the device built on SiO_2 grating. The input-output characteristics of lasers are shown in Figure 5.5(a) and the laser threshold fluence values are also listed in Table 5.2. A low contrast in refractive indices between the substrate and the active medium is not favorable for waveguide mode confinement, so the laser device built on a SU-8 grating substrate (with a high refractive index) delivered a laser emission with a broader spectral line width (FWHM value of 0.6 nm) as well as a higher laser threshold than other devices. The higher slope efficiencies of laser devices built on SiO_2 and glass gratings (both are silica gratings) are advantageous for a potential high laser output power.

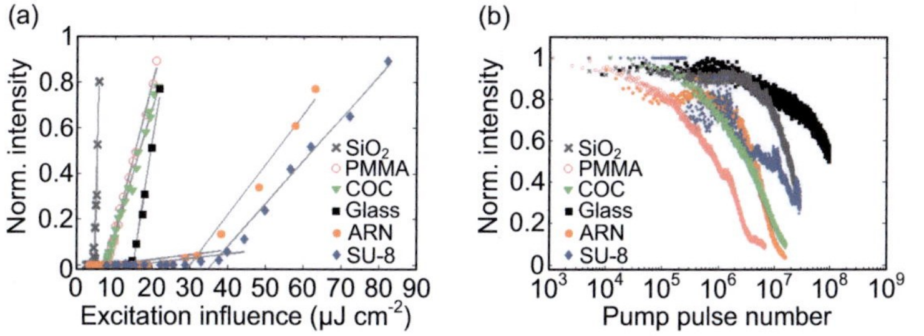

Figure 5.5: (a) Input-output characteristics of organic DFB lasers built with a 220-nm-thick Alq_3:DCM layer on different grating substrates. (b) Degradation characteristics of Alq_3:DCM DFB lasers built on different grating substrates. Pump spot size in all measurements: 550 μm × 450 μm (2×10^{-3} cm^2).

The laser degradation characteristics were monitored under a UV-pump energy of three times the individual laser threshold. Figure 5.5(b) presents the degradation characteristics of the laser peak intensity and the resulting laser lifetime values are listed in Table 5.2 according to the 50 %-drop criterion. The devices built on the glass substrate provided a significantly improved laser emission stability. It is obvious that the devices built on polymer and photoresist grating substrates undergo a higher photodegradation rate. The supplementary photodegradation can be attributed to the chemical instability of polymeric substrate materials in contact with organic gain media. The influence is becoming more dominant along with higher pump pulse energy or longer storage time after laser fabrication. Additionally, for SU-8 and SX AR-N 4800/16 photoresist gratings, the overlap of absorption spectrum with the pump laser spectrum at 355 nm may induce a slight change in the waveguide refractive index and an unstable chemical structure during a long-term measurement. Although the PMMA, COC, SU-8 and SX AR-N 4800/16 grating substrates yielded reasonable laser threshold values and demonstrated narrow laser emission spectra lines, they are not as qualified compared to silica gratings for Raman spectroscopy applications.

Summarizing the measurement results of laser devices on various grating substrates, the silica gratings were preferred for Raman experiments compared to PMMA, COC, SU-8 and SX AR-N 4800/16 grating substrates. Taking advantage of the good transparency, the glass grating substrate could be applied in the excitation path more flexibly than the SiO_2 grating on a opaque thermally oxidized wafer substrate. Therefore, for further work, the D 263® glass gratings were chosen as the DFB-corrugation substrates to fabricate organic laser devices for Raman experiments.

1D- and 2D-DFB-corrugation substrates

As elucidated in Section 2.3.2, using two-dimensional DFB nanostructure could give rise to a reduced laser threshold and an enhanced slope efficiency. As a potential approach to improve the laser performance, it is necessary to compare laser beam profiles and laser emission spectra of organic laser devices built on substrates with one-dimensional (1D) and two-dimensional (2D) DFB-corrugation features.

A 1D-sinusoidal and a 2D-sinusoidal SX AR-N 4800/16 grating substrates with period of $\Lambda_x (= \Lambda_y) = 400$ nm were fabricated using laser interference lithography. The grating depth was chosen to be ~ 40 nm. Figure 5.6 represents the AFM images of 1D- and 2D-DFB-corrugations. The thermal evaporation of 220-nm-thick Alq_3:DCM direct on the photoresist grating substrates finished the fabrication of the laser devices.

Figure 5.7 demonstrates the laser emission spectra and laser threshold curves of 1D-DFB and 2D-DFB lasers. Though the 2D-DFB laser deliv-

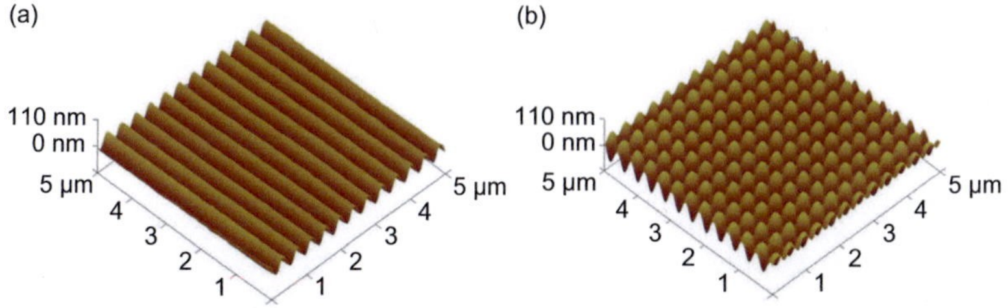

Figure 5.6: AFM images of (a) 1D- and (b) 2D-SX AR-N 4800/16 grating substrates.

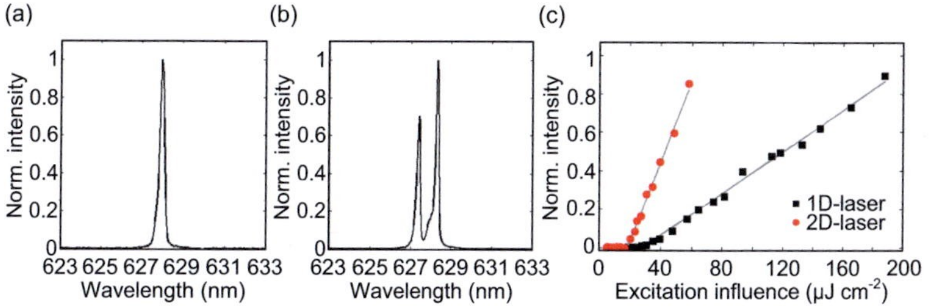

Figure 5.7: Laser spectra of (a) 1D-DFB laser with a peak wavelength at 628 nm and (b) 2D-DFB laser with a main peak wavelength of 628.4 nm and a weaker neighboring peak at 627.4 nm. (c) Input-output characteristics of the 1D and 2D organic DFB lasers with threshold fluences of 32 µJ cm^{-2} at a laser emission wavelength of λ = 628 nm and 20 µJ cm^{-2} at λ = 628.4 nm, respectively. Pump spot size: 550 µm × 450 µm (2 × 10^{-3} cm^2).

ered a lower threshold fluence of 20 µJ cm^{-2} and a higher slope efficiency of 2 % compared to the 1D-DFB laser counterparts (threshold fluence of 32 µJ cm^{-2} and slope efficiency of 0.5 %), the 2D-DFB laser spectrum implied a neighbor longitudinal mode with emission peak at 627.4 nm, as shown in Figure 5.7(b). This characteristic emission spectrum was most likely induced by the deviation in corrugation periods orthogonal to each other (mainly due to the measurement uncertainty for rotating angle). Even if the deviation between corrugation periods was controlled as small as

$\sim$ 2 nm, it led to a multi-longitudinal-mode laser emission, which is unfavorable as excitation source for spectroscopic applications.

Figure 5.8 illustrates the far-field emission patterns of 1D-DFB and 2D-DFB lasers with increasing excitation fluences. For the 1D-DFB laser, a fan-like laser emission was observed due to the inherent lack of lateral confinement. For the 2D-DFB laser, when the pump energy was just over the laser threshold, the far-field pattern indicated an ideal diffraction-limited emission. However, at approximately twice of the threshold fluence ($50\,\mu J\ cm^{-2}$) two perpendicular laser fans were superimposed on the central low divergence spot. With increasing excitation energy, the divergence of horizontal and vertical emission lines extended as well. The overall divergence of the 2D-DFB laser at an excitation fluence of $200\,\mu J\ cm^{-2}$ was almost comparable to that of 1D-DFB laser. As mentioned previously, for Raman spectroscopy applications, the organic laser devices were usually pumped at a high pulse energy level to ensure a sufficient average output power (e.g., $1.6\times10^3\ \mu J\ cm^{-2}$ at a pump spot size of $2\times10^{-3}\ cm^2$). At this pump level, 2D-DFB laser demonstrated no great advantage over 1D-DFB laser in respect of laser emission divergence.

5.2.3 Optimum optical pumping

As elucidated in Section 2.2.3, organic lasers are optically pumped with a short pulse duration to avoid triplet accumulation. A DPSS laser (Crylas, FTSS355-Q2) laser was chosen as the optical pump source for organic laser devices in Raman investigations. To avoid reflections on the glass lid after encapsulation, the organic devices were perpendicularly pumped by this UV-laser [200]. The related parameters of optical pumping include pump repetition rate, pump pulse energy and pump spot size. Below the latter two influence factors are discussed.

To ensure a sufficient laser output power, the Alq_3:DCM laser was pumped with the highest capability of the Crylas DPSS laser: a repetition rate of 10 kHz and a pump pulse energy of $3.2\,\mu J$. At this pump level, the laser line width was investigated through measuring FWHM values of laser spectra at

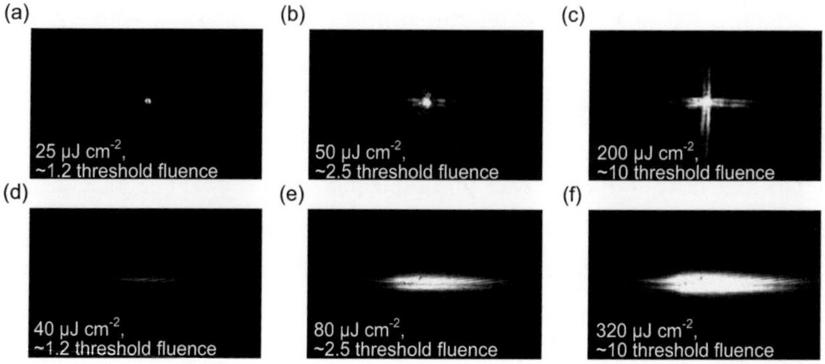

Figure 5.8: Evolution of the far-field emission pattern of the 2D-DFB (a)-(c) and 1D-DFB lasers (d)-(f) (the laser emission line is parallel to the DFB grating orientation) with increasing excitation fluence. The images are the photographs of bottom-emission patterns taken from a camera placed behind lasers.

different pump pulse energies by a spectrometer (Acton, SpectraPro 300i, grating 1800 g mm^{-1}) connected to an ICCD-camera (Princeton Research, PiMax 512) with an optical resolution of ~ 0.1 nm. For an organic laser device with a 210-nm-thick Alq_3:DCM film built on a glass grating substrate (Λ = 400 nm), the FWHMs of laser spectra with the maximum at 629.3 nm grew from 0.1 nm to 0.3 nm in shorter-wavelength direction, when the UV pump pulse enhanced from 60 nJ to 1.29 µJ at a pump area of 550 µm × 450 µm (2×10^{-3} cm^2), as shown in Figure 5.9.

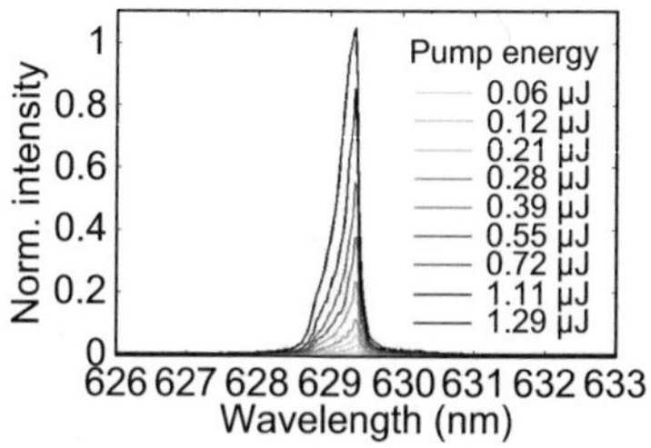

Figure 5.9: Laser spectra measured at different pump pulse energies.

This result verifies the conclusion that the high pump pulse energy facilitates more lateral modes, which increase the spectral width of the laser line [116].

The pump spot size is also determining the laser lifetime and the laser threshold. It has been verified that due to the reduced excitation density the laser threshold (in units of nanojoule per pulse) increases with enlarging the pump spot size. For better comparison, the threshold pulse energy is often normalized to the excitation area (threshold fluence, in units of microjoule per cm^2). Calzado et al. reported that the excitation area for reasons of comparability is only valid if the pump spot area of optical excitation is sufficiently large [291]. The DFB-OSLs can be realized through different fabrication methods in a large number of variants. The different corrugation schemes will result in different coupling mechanisms. The devices have in general neither pure index coupling nor pure gain coupling. Both coupling mechanisms contribute to the laser mechanism and the theoretical treatment becomes complex. To investigate the influence of pump spot size on the laser threshold fluence, a general upper value of pump spot size was derived from different DFB laser configurations to compare the laser threshold fluences of second-order DFB laser devices.

First a DFB-OSL was fabricated using the established thermal evaporation method. A 350-nm-thick layer of Alq_3:DCM was deposited onto a SiO_2 grating substrate with a grating period of 400 nm and a grating height of 90 nm. The AFM micrographs in Figure 5.10(a) show the surface corrugation after deposition which has a good accordance with the original grating pattern. The laser thresholds of the devices were measured at different pump spot sizes and the normalized laser threshold fluences are shown in Figure 5.10(d). For pump spot areas larger than 3×10^{-4} cm^2 the laser threshold fluence leveled off to ~ 10 % of the threshold value for the tightest focus of $\sim 1 \times 10^{-5}$ cm^2. With further enlarged pump spot areas, the threshold fluence decreased only insignificantly. The attenuation rate of threshold fluence along the pump spot area became ~ 0.5 % of the value for tight focuses.

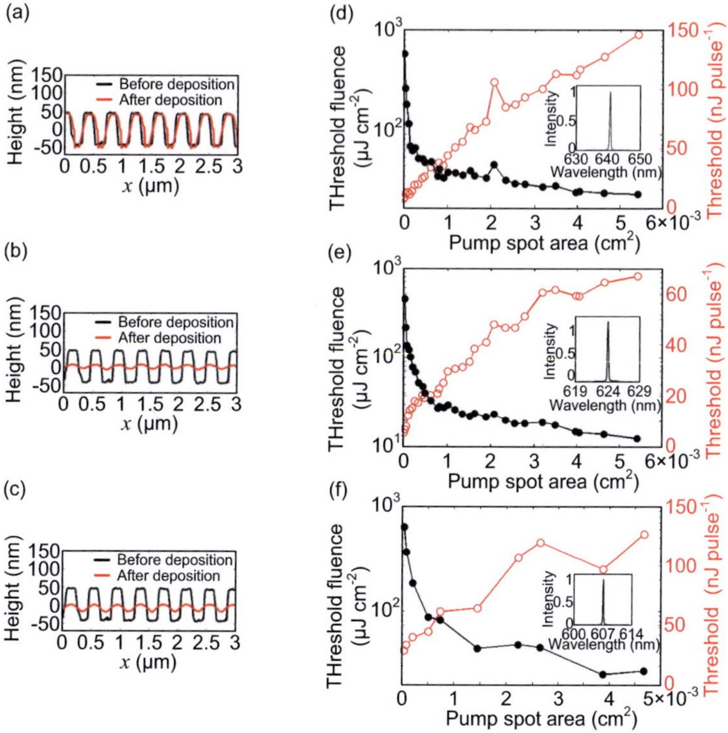

Figure 5.10: AFM micrographs of two surface corrugation patterns before and after (a) thermal evaporation of 350-nm-thick Alq_3:DCM layer, (b) spin-coating of 270-nm-thick $F8_{0.9}BT_{0.1}$:MEH-PPV layer and (c) horizontal dipping of 210-nm-thick $F8_{0.9}BT_{0.1}$:MEH-PPV layer. (d)-(f) Laser threshold fluences and threshold pump energy varying with the pump spot area of corresponding laser devices. Inset: laser spectra with peak wavelengths of 640.6 nm, 623.6 nm and 607.6 nm, respectively.

Besides small molecule DFB lasers, the conjugated polymer DFB lasers were also investigated. To ensure the comparability between various devices, the same SiO_2 grating substrate was utilized as used for the Alq_3:DCM DFB laser. $F8_{0.9}BT_{0.1}$:MEH-PPV was dissolved in toluene at a concentration of 20 mg ml^{-1} (85:15 wt%). Two different solution processing routes, spin-coating and horizontal dipping, have been adopted to deposit an active $F8_{0.9}BT_{0.1}$:MEH-PPV layer with film thicknesses of 270 nm and 210 nm, respectively, onto the grating substrate. The AFM micrographs in Figure 5.10(b) and 5.10(c) show the surface corrugation after deposition. The average corrugation height left after deposition was measured as 10 nm for spin-coating and 20 nm for horizontal dipping. In contrast to Alq_3:DCM DFB lasers fabricated via thermal evaporation, the solution-processed organic DFB lasers revealed a different grating pattern at the air/organic interface from that at the organic/substrate interface. The surface grating height was related to the original grating height and film thickness after deposition. It was found that the surface corrugation height decreased as the film thickness grew. Above a film thickness of 500 nm, the perturbation was lower than 5 % of the original grating height. As shown in Figure 5.10(e) and 5.10(f), the laser thresholds were measured at different excitation spot sizes. The laser threshold fluences followed the same tendency as that described above for the Alq_3:DCM laser device. It was noticed that above a pump area of about 1×10^{-3} cm^2, the laser threshold fluence decreased to 7 % of the initial value for the tightest focus ($\sim 1 \times 10^{-5}$ cm^2).

The laser threshold fluences were compared for three different DFB-OSL devices fabricated via thermal evaporation, spin-coating and horizontal dipping. Figure 5.11 summarizes the comparison and shows a similar tendency for all three different corrugation configurations. The laser threshold fluence decreases with growing excitation area and coupling strength $\kappa_{\mathrm{eff}}L$, where κ_{eff} is the coupling coefficient of two counter-propagating fundamental waveguide modes and L is the effective device length. The saturation-like behavior of the threshold fluence can be reproduced above a certain circular pump spot area. Similar to the prediction made by Kogelnik et al. [175], when the pump spot is too small and the coupling strength $\kappa_{\mathrm{eff}}L < 1$, the device is excited at a state of "undercoupling". Due to the deficiency of coupling, the laser threshold is much higher compared to the sufficient coupling strength. Conversely, when the pump spot is increased to another limit, in this case $\kappa_{\mathrm{eff}}L \geq 5$, the device will be excited at a state of "overcoupling". Hence, the laser threshold influence cannot be further decreased and reaches a saturation value.

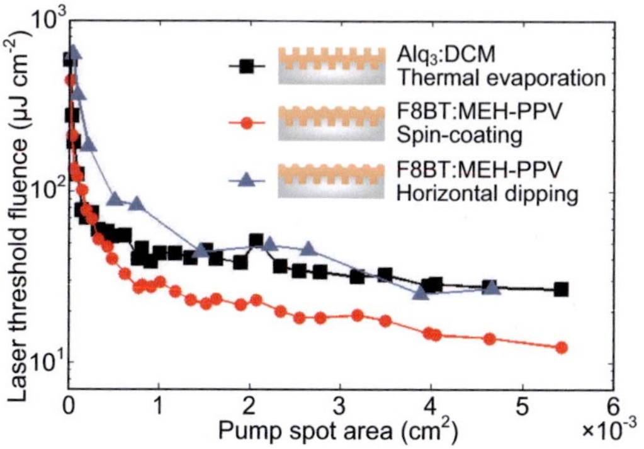

Figure 5.11: Laser threshold fluences for varying pump spot areas for various film thickness modulation configurations fabricated via thermal evaporation, spin-coating and horizontal dipping.

The measurement results gave hints about the laser threshold fluence and the pump spot size dependency for organic DFB lasers in general. Nevertheless, differences on the level-off pump spot sizes were found in three devices fabricated via different routes. The laser threshold fluence is closely related to the coupling coefficient κ_{eff} and the effective device length L. The coupling coefficient is not only related to the surface configuration, but also influenced by the absorption coefficient and the gain coefficient of employed active media. With a higher coupling coefficient, overcoupling can be reached for a smaller pumped area. In addition, different grating periods will also change the illuminated grating line number contributing to the laser oscillation. Thus, the pump spot areas, above which the threshold fluence is comparable for different devices (the attenuation rate of threshold fluence decreases to ~ 0.5 % of the value for tight focuses), are not the same [275]. For Alq_3:DCM laser fabricated by thermal evaporation this value was 3×10^{-4} cm^2; the other values are 6×10^{-4} cm^2 for $F8_{0.9}BT_{0.1}$:MEH-PPV laser fabricated by spin-coating, and 1.5×10^{-3} cm^2

for $F8_{0.9}BT_{0.1}$:MEH-PPV laser fabricated by horizontal dipping, respectively. The upper criterion spot size 1×10^{-3} cm^2 for invariant laser threshold fluence was also found in the work of Calzado, et al. [291].

Conclusively, a typical pump spot size of 2×10^{-3} cm^2 was applied for the Alq_3:DCM lasers in Raman experiments.

5.2.4 Encapsulation

The characterization of organic lasers was accomplished in a vacuum chamber ($< 5 \times 10^{-3}$ Pa) to protect the active medium from photooxidation. However, as an excitation source for Raman scattering measurements the organic laser device stands often alone in the ambient air, either for free-beam or fiber-coupled configuration. Therefore an encapsulation against water and oxygen is required. The encapsulation was completed in nitrogen environment by bonding two glass lids with an ultraviolet (UV) curable optical adhesive (Norland, NOA68) enclosing an organic laser sample in between, as depicted in Figure 5.12. Thus, the bottom-emission from

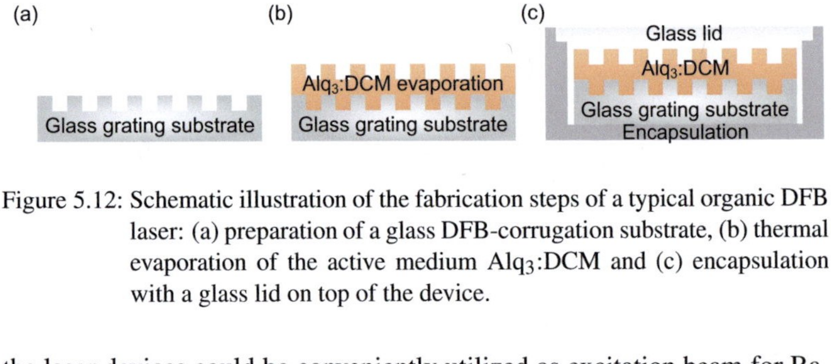

Figure 5.12: Schematic illustration of the fabrication steps of a typical organic DFB laser: (a) preparation of a glass DFB-corrugation substrate, (b) thermal evaporation of the active medium Alq_3:DCM and (c) encapsulation with a glass lid on top of the device.

the laser devices could be conveniently utilized as excitation beam for Raman applications. After encapsulation the laser emission wavelength and the laser threshold of organic laser device stayed at the values measured in the vacuum chamber. The laser degradation characteristics were also comparable to the results measured in the vacuum chamber. As shown in Figure 5.13, the organic laser after encapsulation demonstrates a laser emission lifetime as long as 9.8×10^7 pulses under optical pumping at two times the laser threshold and a repetition rate of 5 kHz, corresponding to a time period of 5.4 hours of continuous operation. During the measurement no shift in laser peak wavelength was observed.

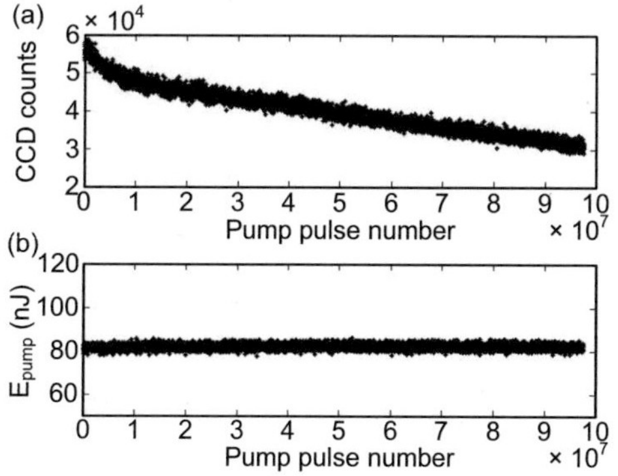

Figure 5.13: (a) Degradation characteristic of an encapsulated Alq_3:DCM laser device and (b) variance of pump pulse energy with the elapsed pump pulse number. Measurement conditions: repetition rate 5 kHz, pump spot size 550 µm × 450 µm (2×10^{-3} cm^2).

5.2.5 Discussion

In this section various measures to improve organic laser properties were introduced from three points of view: organic active medium, DFB-corrugation substrate and optical pumping. In most cases, the laser emission characteristics are influenced by multiple aspects. A compromise among different requirements plays an important role in determining the optimum laser design. For example, to ensure a higher laser emission stability, the plan to employ an even higher pump energy and pump repetition rate has to be abandoned. Meanwhile, at a high pump pulse energy level, the laser line width gets broader unavoidably. It is necessary to find a balance point in consideration of adequate output power, high emission stability, high spatial mode quality and moderately narrow line width. It turns out that an encapsulated Alq_3:DCM organic laser built on a qualified 1D glass grating is the final choice for Raman experiments.

5.3 Raman experiments in free-beam coupling configuration

In this section the organic semiconductor DFB lasers are applied as excitation sources in Raman measurements using free-beam coupling configuration. The Raman spectra of sulfur (S_8) and cadmium sulfide (CdS) excited by DFB-OSLs are demonstrated as examples. In the free-beam coupling configuration, the organic laser emission after collimation is directly guided into a confocal microscope to excite the Raman spectra of the analytes. This ensures a high excitation power on the analytes with a minimum coupling loss.

5.3.1 Optical setup for free-beam Raman excitation

As discussed previously, the organic semiconductor Alq_3 doped with the laser dye DCM forms a very efficient and stable Förster energy transfer system and was chosen as the active medium for the laser devices. To match the Alq_3:DCM red emitter, a glass grating (Visolas GmbH) with a 1D-corrugation period of 400 nm was applied as the second-order distributed feedback resonator. Alq_3 and 2.3 % by weight of DCM were thermally co-evaporated in a high vacuum chamber with a thickness of 210 nm onto the glass grating substrate. An encapsulation was achieved by bonding a glass lid through ultraviolet curable optical adhesive (Norland, NOA68). Figure 5.14(a) shows the encapsulated DFB-OSL when vertically pumped on an area of $550\,\mu m \times 450\,\mu m$ (2×10^{-3} cm^2) by a DPSS laser (Crylas, FTSS355-Q2) with a wavelength of 355 nm. The bottom emission of the laser device was utilized for a convenient coupling of the laser beam into the Raman confocal microscope. The laser line width was investigated through measuring the FWHMs of the laser spectra at different pump pulse energies. As presented in Figure 5.9, the FWHMs of the laser spectra with the maximum at 629.3 nm grew from 0.1 nm to 0.3 nm, when the UV pump pulse energy increased from 60 nJ to 1.29 μJ. This behavior was attributed to higher lateral modes which were excited for higher pump pulse levels. The DFB-OSL exhibited a slope efficiency of 7.6 % and a laser threshold fluence of 15.9 μJ cm^{-2}, as shown in Figure 5.14(b).

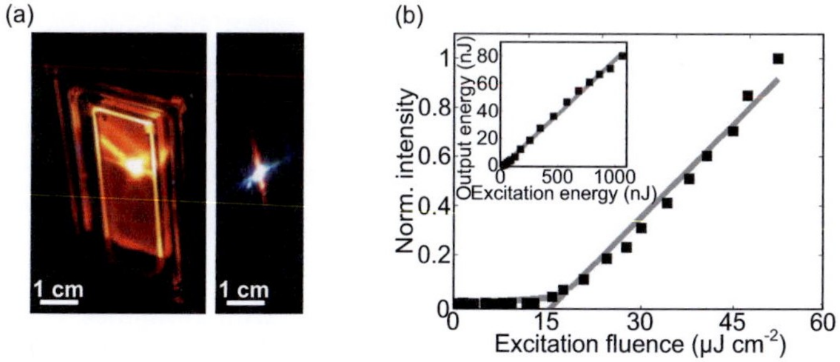

Figure 5.14: (a) Photograph of an encapsulated DFB-OSL when pumped by the ultraviolet pump laser (left) and laser emission beam profile before coupling into the setup for Raman measurement (right: red organic laser emission line and white-blue UV pump laser spot). (c) Input-output characteristic at a wavelength of 629.3 nm for lower pump energy level and higher pump energy level (inset).

For Raman scattering measurements, the UV pump pulse energy was adjusted during the operation through a variable neutral density filter and controlled by a pulse energy meter (Coherent, LabMax-TOP). The maximum pump pulse energy was measured as 3.2 μJ. Under this pumping condition, the encapsulated DFB-OSL achieved a maximum output pulse energy of 160 nJ, which corresponds to an averaged laser power of 780 μW at a pump repetition rate of 10 kHz. The FWHM of the laser spectrum was measured to be 0.5 nm.

The setup for Raman spectroscopy is shown in Figure 5.15(a). It consists of a UV pump laser, an encapsulated DFB-OSL, an inverted confocal microscope (Nikon, Eclipse TE2000-U) and a detection system. A clean-up bandpass filter (Semrock, F94-633L) with a bandwidth of 2.4 nm at the entrance of the confocal microscope was inserted to suppress ambient light and photoluminescence of the organic active medium. Making use of a 45°-tilted dichroic beam splitter (Semrock, F78-630), the organic laser emission was guided to excite the Raman scattering of a sample on a horizontal translation stage. Compared to more conventional lasers, e.g., a helium-neon (He-Ne) laser, the DFB-OSL laser emission is perpendicular to the substrate as a slightly divergent fan parallel to the orientation of the grating grooves as shown in Figure 5.14(a). Applying a collimation lens and a 40× objective with a numerical aperture of 0.6, the emitted laser stripe could be focused to a length of less than 100 μm on the sample as shown in the inset of Figure 5.15(b). The laser power on the sample reached up to 350 μW. A longpass filter (Semrock, F76-633) was employed in the setup to further suppress the laser reflection, the Rayleigh scattering of the analytes and the neighboring amplified spontaneous emission (ASE) of the organic active medium. The collected Raman signals were analyzed and recorded by a spectrograph (Horiba, Triax 320, grating 300 g mm^{-1}) connected to a liquid-nitrogen-cooled CCD-Camera (Horiba, Symphony LN2 detector). This combination of the detection system gave a spectral resolution of 0.26 nm, corresponding to a Raman shift resolution of 9.2 cm^{-1} for the excitation wavelength of approximately 630 nm.

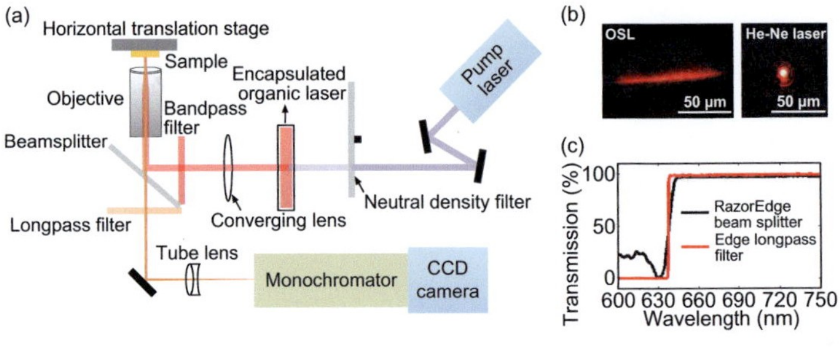

Figure 5.15: (a) Schematic illustration of the setup for Raman spectroscopy. (b) Laser beam profiles of DFB-OSL and He-Ne laser at the same sample plane. (c) Transmission spectra of the dichroic (RazorEdge) beam splitter and the longpass filter.

The laser degradation characteristics of the DFB-OSLs were discussed in Section 5.1.3 and 5.2. For Raman experiments, the optical pump pulse energy was kept at a high level to ensure a sufficient organic laser emission power. If pumped at a pulse energy of 1.8 µJ, the laser power decreased only 5 % after 1.8×10^7 pump pulses, corresponding to 30 min at a repetition rate of 10 kHz. At a pump pulse energy of 3.2 µJ, the laser emission from one pumped spot decayed 10 % within 30 min, as shown in Figure 5.16. This stability of one single pump spot is sufficient for Raman measurements lasting few seconds to few minutes. Note that the initial DFB-OSL power could be completely recovered by slightly moving the device to another pump spot position on the active layer.

The pump spot size plays an important role in determining the laser emission lifetime. Figure 5.16 presents the comparison result of laser degradation characteristics at two different pump spot sizes. Pumped by a smaller spot size of 200 µm × 150 µm (0.2×10^{-3} cm^2), which was achieved by using a focusing lens, the initial instantaneous laser output power was higher, but quickly leveled off due to the higher degradation rate. In comparison, at a larger pump spot size at 550 µm × 450 µm (2×10^{-3} cm^2), a compromise between a moderate output power and a high laser emission stability was ensured.

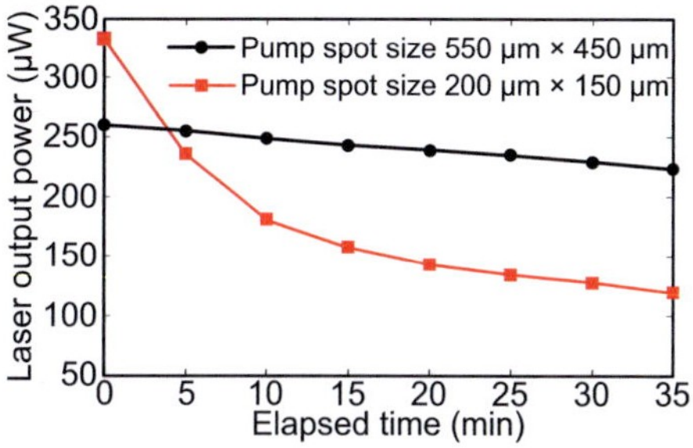

Figure 5.16: Degradation characteristics of the encapsulated Alq_3:DCM DFB laser under pump pulse energy of 3.2 µJ and repetition rate of 10 kHz at two different pump spot sizes.

5.3.2 Raman measurements

Sulfur (S_8) was applied as the first analyte in this work for Raman experiments, which is a strong Raman scatterer with dominant bands at $\sim$ 219 cm^{-1} and $\sim$ 473 cm^{-1} [314, 315]. Currently, Raman spectroscopy for sulfur and sulfur compounds attracts a lot of interest for in-situ detection of marine organisms and minerals influenced by sulfidic environments [316, 317, 318]. Using an integration time of 30 s for spectra acquisition, Raman spectra of the sulfur powder sample (Sigma-Aldrich, $>$ 99.5 %) were acquired, which were excited by the DFB-OSL with a laser emission peaked at 629.3 nm. The Raman peaks at 219 cm^{-1} and 473 cm^{-1} could be easily identified from the recorded Raman spectra, as shown in Figure 5.17(a). Along with the increased excitation power, the Raman sig-

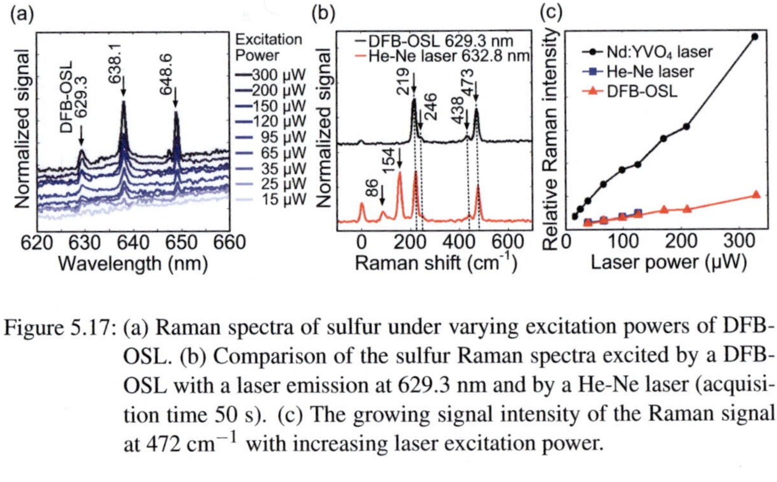

Figure 5.17: (a) Raman spectra of sulfur under varying excitation powers of DFB-OSL. (b) Comparison of the sulfur Raman spectra excited by a DFB-OSL with a laser emission at 629.3 nm and by a He-Ne laser (acquisition time 50 s). (c) The growing signal intensity of the Raman signal at 472 cm^{-1} with increasing laser excitation power.

nal intensity grew as well and was linearly proportional to the laser power. Above an excitation power of 120 μW the weak Raman band at 438 cm^{-1} could also be identified. Compared to the sulfur Raman spectra excited by the He-Ne laser (Coherent, Model 200 single frequency stabilized HeNe laser) with a random polarization, a good correspondence was found in the Raman peaks at 219 cm^{-1}, 246 cm^{-1}, 438 cm^{-1} and 472 cm^{-1}. The polarization of the excitation source has no influence on the Raman lines of sulfur due to the symmetry of the orthorhombic crystal [315]. As shown in Figure 5.17(b), the width of the Raman lines is comparable to those excited by the He-Ne laser. The missing peaks at 154 cm^{-1} and 86 cm^{-1} in the excited Raman spectrum could be explained by the filter function of the commercial beam splitter and longpass filter, which are designed for a standard He-Ne laser with an excitation wavelength of 632.8 nm. For the organic laser with an excitation wavelength of 629.3 nm, the excited Raman shifts smaller than 167 cm^{-1} overlap with the attenuation range of the filter system ($\geq$ 636 nm, see Figure 5.15(c)).

To further verify the feasibility of DFB-OSLs as excitation sources for Raman scattering measurements, the intensities of sulfur Raman signal at $\sim$ 472 cm^{-1} were compared, which were excited by a diode pumped, neodymium-doped yttrium aluminum garnet (Nd:YAG) laser with a wavelength of 532 nm (HB-Laserkomponenten), a He-Ne laser with a wavelength of 632.8 nm and a DFB-OSL with a wavelength of 629.3 nm. Figure 5.17(c) shows the growing Raman signal along with the increased excitation power under an acquisition time of 150 s. The Raman signal intensity excited by the Nd:YAG laser was approximately four times higher than the one excited by the DFB-OSL. This can be partly explained by the fact that the Raman scattering intensity is inversely proportional to λ^4 [218]. The Raman signal intensity excited by the DFB-OSL were comparable to that excited by the He-Ne laser at the same excitation power. The results showed that the DFB-OSL is qualified to replace a He-Ne laser in Raman scattering measurements.

To match commercially available optical filters designed for the use with He-Ne lasers at 632.8 nm, one DFB-OSL with a close emission wavelength was required. It is possible, but quite elaborate to evaporate an organic active layer with the necessary precision within a few nanometers. Therefore, a continuously tunable DFB-OSL was applied in this Raman measurements. This laser was fabricated by evaporating a wedge-shaped active medium layer onto the glass grating substrate. The fabrication details can be found in Section 3.2.1. Attributed to the correlation between active layer thickness and DFB-OSL emission wavelength, a specific laser emission wavelength can be obtained by pumping at a spatially defined position on the wedged layer. Figure 5.18(a) shows a photograph of the laser sample built on a glass grating substrate with a corrugation period of 400 nm after coating with a wedge-shaped film of Alq_3:DCM. It revealed a spatially rendering color which emerged from wedge-shaped thin films interference effects. The evaporated film thickness changed linearly from 190 nm to 380 nm as measured on a reference sample. The laser sample was scanned with a spatial resolution of 100 µm and a spectrum was taken at each point. The laser emissions varied from 617 nm to 653 nm continuously, as shown in Figure 5.18(b). Using this approach the laser emission was tuned exactly to a wavelength of 633 nm in order to match the commercial optical filters, as shown in Figure 5.18(c).

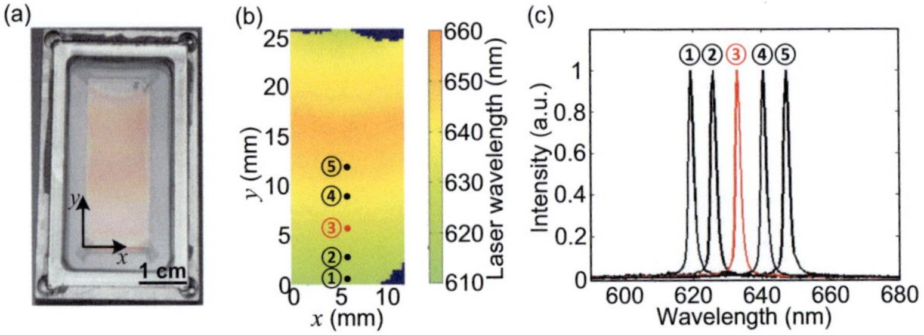

Figure 5.18: (a) Photograph of an encapsulated spectrally tunable DFB-OSL with a continuously changing film thickness of Alq_3:DCM fabricated by rotating shadow mask evaporation. (b) Spatially resolved laser wavelengths of the laser sample. (c) Laser spectra measured at different pump spot positions ①-⑤ on the wedge-shaped DFB laser. Among them, the peak wavelength of the laser spectrum from position ③ is exactly at 633 nm.

For an acquisition time of 50 s the Raman peaks at 153 cm^{-1} and 81 cm^{-1} turned up and the Raman intensities were comparable with the ones excited by the He-Ne laser, as shown in Figure 5.19. Only slight differences could be observed regarding the width of the Raman lines. Since the pump area on the organic laser sample (2×10^{-3} cm^2) for the tunable laser covered a slightly wedge-shaped organic layer, the laser line width grew to 0.6 nm for an UV excitation pulse energy of 3.2 µJ. In turn, this led to slightly broader lines than in the case of excitation with the He-Ne laser.

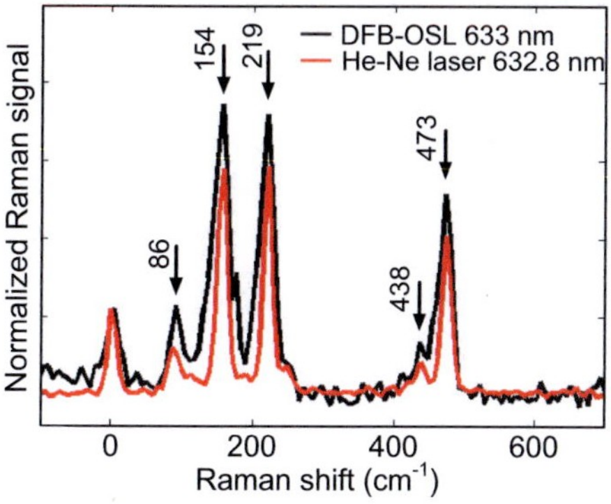

Figure 5.19: Comparison of sulfur Raman spectra excited by the free organic DFB laser beam with a laser wavelength of 633 nm and the He-Ne laser. Measurement conditions: 20× objective, 300 g mm^{-1} grating, 0.3 mW excitation intensity, 50 s acquisition time in both cases.

The DFB-OSLs were also employed to investigate the Raman spectra of other analytes. Figure 5.20 shows the Raman lines of cadmium sulfide (Crystal GmbH) excited by the non-wedged DFB-OSL with a laser emission at 629.3 nm. Two prominent Raman peaks at 307 cm^{-1} and 600 cm^{-1} were identified, which originated from the CdS longitudinal optical (LO) phonon mode and the second longitudinal optical (2LO) phonon mode, respectively [320, 321]. In addition the multiphonon scattering modes at 216 cm^{-1}, 249 cm^{-1}, 327 cm^{-1} and 346 cm^{-1} were observed as well [322]. The Raman spectrum was comparable with the one excited by the He-Ne laser and matched well with the Raman database of the unoriented bulk wurtzite-type crystals of CdS [319].

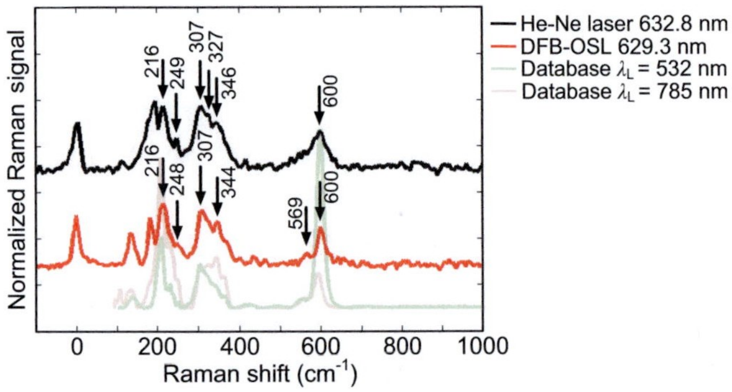

Figure 5.20: Raman spectra of cadmium sulfide excited by the DFB-OSL with a laser wavelength of 629.3 nm and the He-Ne laser in comparison with the spectra from Raman database [319].

5.4 Raman experiments in fiber-coupling configuration

In addition to using the free-beam configuration, the laser can be coupled into an optical fiber to achieve a modular laboratory Raman system. Due to the absent confinement in the direction of grating orientation, the laser radiation of devices built on one-dimensional grating substrates is emitted in a fan-like shape, which increases the difficulty of laser focusing and collimation. It may cause a high coupling loss to couple DFB-OSL into a single-mode fiber due to the inherent lateral laser modes. Nevertheless, a flexible and convenient modular system opens up the application fields of organic laser. The fiber coupling separates the Raman microscope and the DFB-OSL into stand-alone modules. This may replace the bulky optical system, which highlights the advantages of miniaturized organic laser devices.

5.4.1 Optical setup for fiber-coupled Raman excitation

Figure 5.21 shows a schematic illustration of the modular Raman system including a stand-alone upright optical microscope, a spectrograph and the

DFB-OSL as excitation source.[7] Different from free-beam coupling, the bottom-emission from the DFB-OSL was utilized in addition to the surface emission. The bottom emission was harvested by adding a reflector film (3M, Vikuiti Enhanced Specular Reflector ESR) on the bottom of the DFB grating substrate, as shown in Figure 5.22(a) and 5.22(b). Thus, the surface- and bottom-emissions were acquired simultaneously and almost doubled the power of DFB-OSL in one direction. Pumped by the UV laser, the DFB-OSL with a homogenous 220-nm-thick active layer emitted at a peak wavelength of 631.4 nm. This DFB-OSL device exhibited a slope efficiency of 6.8 % and a laser threshold fluence of 28.4 μJ cm^{-2} at a pump area of 2×10^{-3} cm^2, as shown in Figure 5.22(c). A dichroic beamsplitter was used to separate the organic laser and UV-pump laser emissions.

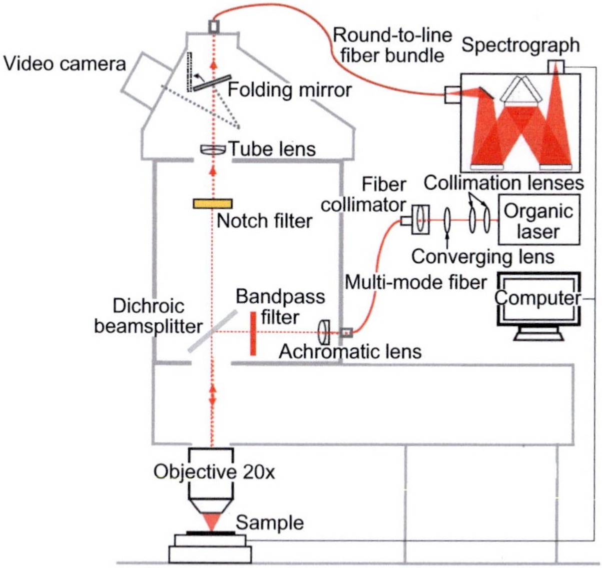

Figure 5.21: Schematic illustration of the overall Raman spectroscopy setup with fiber-coupled organic laser excitation.

[7]The modular Raman system was built at the Institute of Nanotechnology (INT-KIT).

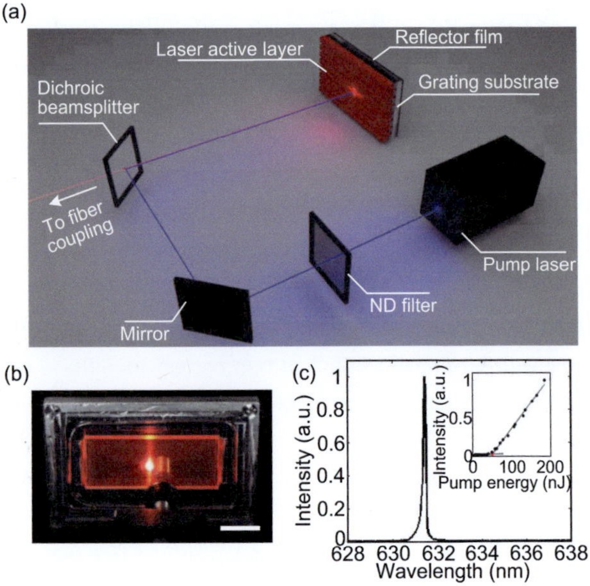

Figure 5.22: (a) Schematic illustration of the organic laser module. (b) Photograph of an encapsulated DFB-OSL when pumped by the UV laser. The scale bar corresponds to 1 cm. (c) Laser spectrum of the DFB-OSL fabricated with a homogenous 220-nm-thick Alq_3:DCM active layer and the input-output characteristic (inset) at the emission wavelength of 631.4 nm.

Due to the lateral laser modes, a coupling loss as high as 88 % was measured when a DFB-OSL was coupled into a single-mode fiber with core diameter of 3 µm. To ensure a high fiber coupling efficiency, multi-mode fibers were used to connect the modules.

The original emission profile from a one-dimensional DFB-OSL has a large divergence angle of over 0.17 rad. To improve the laser divergence, three lenses were applied to realize a well-collimated and focused laser radiation. After improvement via the first two collimation lenses, the divergence angle of the DFB-OSL device was reduced to 13.6 mrad. Further through a converging lens with a focal length of 25 mm and a collimator on the fiber port, the DFB-OSL beam with a final divergence angle of 11.8 mrad was delivered horizontally by a multi-mode fiber with a core diameter of 50 µm (Thorlabs, M42L02) into the confocal imaging Raman microscope (WITec, CRM200). Through a clean-up bandpass filter (Semrock, F94-633L) the laser excitation beam was diverted down to an objective lens (Nikon, Plan, 20×, NA = 0.40) by a dichroic beamsplitter (Semrock, F78-630). The OSL excitation spot diameter on the sample surface was about 20 µm.

In this system, the achromatic lens, the clean-up bandpass filter and the dichroic beamsplitter are integrated in one optical module. Thus, in case of exchanging the excitation source, the setup modification could be easily achieved by changing to a new optical module with an appropriate dichroic beamsplitter and a suitable notch filter. The backscattered Raman light was collected by the objective, filtered by a beamsplitter and a notch filter, focused onto a multi-mode optical fiber, finally guided into the spectrograph (Acton, SP2300i) and captured by a back-illuminated CCD-camera operating at −65 °C (Andor, DV401A).

The fiber coupling efficiency was measured to be 70 %. Considering the optical loss due to collimation optics, fiber coupling, beamsplitter and objective, the overall coupling efficiency was determined to be 40 %. The average DFB-OSL excitation power on the analytes could reach up to 250 µW.

5.4.2 Raman measurements

A multi-mode fiber with a core diameter of 50 µm was used in the excitation path to achieve a high coupling efficiency. The back-scattering light volume was about 20 times larger than the volume excited with a single-mode fiber with core diameter of 3 µm. Due to the enlarged signal beam on the exit of the microscope, a multi-mode fiber with a larger core diameter was used in the detection path to ensure a high Raman signal intensity transferred to the spectrograph. Figure 5.23 shows the Raman spectra of sulfur excited by the DFB-OSL and measured with fibers in the detection path with core diameters of 50 µm, 100 µm and 200 µm, respectively. Al-

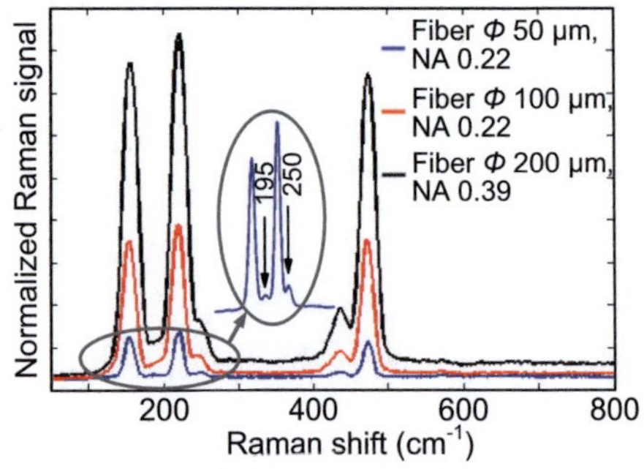

Figure 5.23: Raman spectra of sulfur excited by a fiber-coupled DFB-OSL with a laser emission at 631.4 nm. The spectra were acquired under the same conditions: 20× objective, 600 g mm^{-1} grating, 0.2 mW excitation intensity and 20 s acquisition time, but with Raman light transferring fibers of different core diameters. Inset: magnified Raman signals measured with a multi-mode fiber with a diameter of 50 µm.

though it shows a significant enhancement in Raman signal intensity using a fiber with larger core diameter, a poorer spectral resolution could not be avoided. The Raman peaks at 195 cm^{-1} and 250 cm^{-1} were not detectable using fibers with diameters of 100 µm and 200 µm.

To solve this problem, a round-to-line fiber-bundle (LightGuideOptics, L2000) was employed to connect the exit port of the microscope and the entrance of the spectrograph. It consists of fifteen 100 µm-thick fibers, which form a round terminal in diameter of 570 µm for the microscope exit port and a narrow line form in dimension of 125 µm × 1875 µm as an entrance slit for the spectrograph, as shown in Figure 5.24(a). The fiber-bundle ensured an efficient light transfer and an acceptable optical resolution in Raman spectroscopy. Figure 5.24(b) shows the comparison of the sulfur Raman spectra measured with a 200 µm-fiber and a round-to-line fiber-bundle. Both, the Raman intensity and the spectral resolution were improved.

To verify the functionality of organic lasers as excitation sources for Raman spectroscopy, the DFB-OSL was also employed to investigate the Raman spectra of other analytes. As shown in Figure 5.25(a) and 5.25(b),

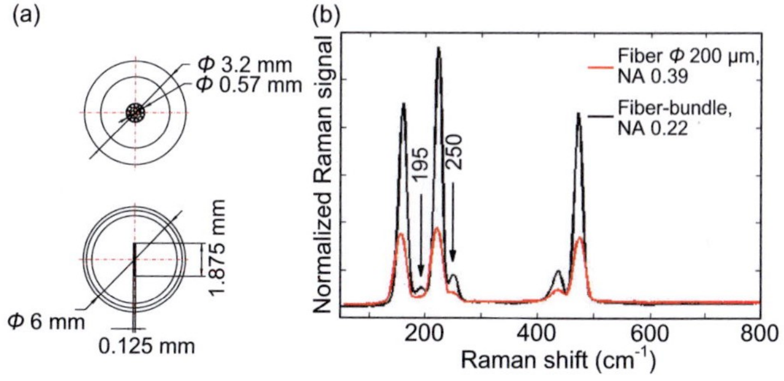

Figure 5.24: (a) Technical drawings of the fiber terminals for the microscope (top) and for the spectrograph (bottom) of the round-to-line fiber-bundle. (b) Sulfur Raman spectra acquired with a standard 200 µm fiber and the round-to-line fiber-bundle. Measurement conditions: 20× objective, 600 g mm^{-1} grating, 0.2 mW excitation intensity, 20 s acquisition time. Due to the improved resolution and throughput, weaker Raman peaks at 195 cm^{-1} and 250 cm^{-1} could be detected with the fiber-bundle.

the Raman spectra of cadmium sulfide and cyclohexane were excited by an organic laser with a power of 200 µW. The intensity and the line width of the main Raman peaks were again comparable with the ones excited by the He-Ne laser under the same excitation conditions.

5.5 Conclusion

Organic semiconductor lasers (OSLs) are good candidates as tunable laser sources available for Raman spectroscopy. In this chapter, OSLs were first introduced as monochromatic excitation sources into Raman experiments using free-space configurations: free-beam coupling and fiber coupling. In addition to the standard requirements of organic lasers in an application-oriented view, i.e., lowering the threshold and extending the wavelength tunability, there are some specific requirements for the application in Raman spectroscopy. In contrary to other application fields of OSLs, e.g., sensing and telecommunication, where the few nanojoule energy level

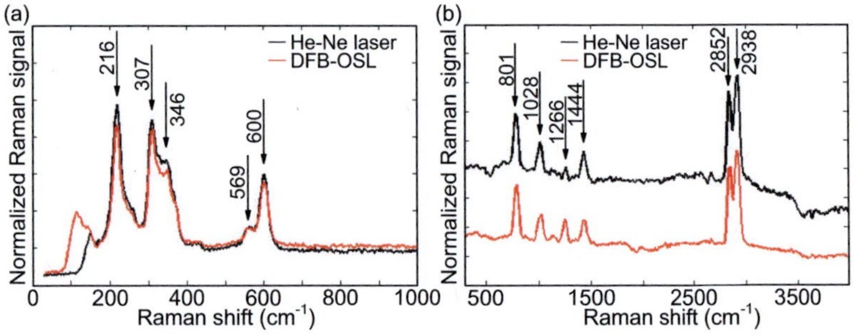

Figure 5.25: (a) Comparison of the cadmium sulfide Raman spectra excited by the fiber-coupled organic DFB laser with a laser wavelength of 631.4 nm and the He-Ne laser. Measurement conditions: 20× objective, 600 g mm^{-1} grating, 0.2 mW excitation intensity, 20 s acquisition time. (b) Comparison of the cyclohexane Raman spectra excited by the fiber-coupled organic DFB laser and the He-Ne laser. Measurement conditions: 20× objective, 150 g mm^{-1} grating, 0.2 mW excitation intensity, 20 s acquisition time.

could be sufficient, a high average output power is desired for Raman analysis. By enhancing the laser power using high optical pump pulse energy and high repetition rate, some issues, e.g., the broadened emission line width, the enlarged beam profiles and the accelerated device degradation should be noted. In this chapter various approaches have been introduced to improve the organic laser properties with regard to three aspects: organic gain medium, DFB-corrugation substrate and optical pumping. It was found that a balance among laser spectral width, laser output power and laser emission stability is the key for Raman spectroscopy applications.

Via employing a qualified glass DFB-grating substrate and an efficient gain medium Alq_3:DCM, the OSLs show a maximum laser slope efficiency of 7.6 % and offered an excitation power up to 350 μW. In the free-beam coupling configuration, the organic laser emission after collimation was directly guided into an inverted confocal microscope to excite the Raman spectra of analytes. This ensured a high excitation power on the analytes with a minimum coupling loss. In comparison, the fiber coupling configuration suffered from a higher coupling loss due to the intrinsic fan-like OSL emission strip, but enabled a flexible modular laboratory Raman system. With help of a multi-mode fiber, the OSL-beam can be guided into an upright Raman microscope with a high coupling efficiency of 70 %. To improve the detection efficiency and the spectra resolution, a round-to-line fiber-bundle was employed to connect the microscope exit port and the spectrograph entrance.

For both Raman setup configurations, DFB-OSLs have been successfully applied as excitation sources in Raman spectra analysis of sulfur, cadmium sulfide and cyclohexane. A tunable organic semiconductor laser was introduced to facilitate a specific laser wavelength to match the commercially available optical filters designed for use with He-Ne lasers at 632.8 nm. Note that the laser line width of tunable DFB-OSLs based on a wedge-shaped active layer is slightly wider than the OSLs built by a uniform organic layer due to the film thickness inhomogeneity. The possible solution is improving the thickness gradient of the active film, e.g., by reducing the slope of the wedge-shaped active layer for a negligible thickness variation within an optical pump spot range.

The large Raman scattering efficiencies of sulfur and cadmium sulfide enabled a straightforward Raman detection. In the case of biomolecular analytes, the lower Raman cross-sections, the limited excitation power of OSLs, and typically lower concentrations of analytes would lead to very small or undetectable Raman signals under comparable excitation/detection conditions. A promising approach for overcoming this challenge is applying surface-enhanced Raman scattering (SERS) effect to amplify the Raman signals, which will be introduced in Chapter 6.

6 Tailored Nanostructures for OSL SERS Measurements

In this chapter novel nanostructures and organic semiconductors are combined for surface-enhanced Raman scattering (SERS) measurements. SERS has the potential to significantly improve the sensitivity of Raman detection using organic semiconductor lasers as excitation sources. After an introduction into state-of-the-art SERS techniques, a new fabrication technique for SERS substrates via laser-assisted replication is described. The third and the fourth subsections demonstrate the functionality of the Au/COC SERS substrates from aspects of computational simulation and experimental results. Finally, an organic semiconductor DFB laser was introduced as an excitation source in the SERS biomolecular analytics.[1]

In general, the integration of high power laser sources into lab-on-a-chip devices is challenging. The integration is not an issue for OSLs which have a typical power level of less than 1 mW. Chapter 5 also demonstrated the application of organic semiconductor DFB lasers (DFB-OSLs) as excitation sources for Raman measurements of sulfur and cadmium sulfide powders. Although a low average power of the OSL ($\sim$ 0.2 mW) was used in those measurements, the large Raman scattering efficiencies of the analytes enabled a straightforward Raman band detection. In the case of biomolecular analytes, the lower Raman cross-sections, the limited excitation power of LOCs, as well as typical low concentrations of analytes would lead to very small or undetectable Raman signals under comparable excitation/detection conditions. A promising approach for overcoming this challenge is to use

[1] Parts of the results have been published in Ref. [323].

the surface-enhanced Raman scattering (SERS) effect which greatly amplifies the Raman signal.

6.1 State of the art

SERS is one of the very few methods which can effectively give specific molecular information about an adsorbate on or near metallic nanostructures. The following main advantages make it very powerful for chemical and biological sensing applications:

1. SERS provides a complementary technique to fluorescence, which is also widely used as an ultra-sensitive and single-molecule spectroscopic tool in biophysics. However, to achieve a large fluorescence quantum yield, the molecular samples have to be labeled by a fluorescent dye. In comparison, SERS does not require any labeling since it is based on the intrinsic vibrational level properties of the molecule.

2. For analysis of aqueous solutions, for instance in life science, Raman detection has an important advantage over infrared or near-infrared absorption, namely, practically no perturbing signal from water [324, 325, 326]. Moreover the detection limits could be decreased by many orders of magnitude by exploiting the SERS effect. The high SERS enhancement factor on the order of $> 10^{10}$ allows detection at the single-molecule level.

3. Because the vibrational relaxation times are much shorter than the fluorescence lifetimes, the number of Raman photons emitted by a molecule per unit time can be higher than that of fluorescence photons by factors $\sim 10^2 - 10^3$ [221]. The high enhancement of many orders of magnitude over spontaneous Raman detection method enables SERS to collect the distinguishable spectral properties within few seconds. This short integration times provide an opportunity for SERS in detecting living cells or Raman mapping for monitoring chemical changes within one cell.

As one of the exciting applications, SERS biomedical sensors are used in detection of various biological samples and diseases, including DNA and gene probes, neurotransmitters, living bacteria, human cells, tumor fragments and so on [224]. In chemistry and material science, SERS is widely used in detecting chemical agents and in the study of electrochemically active molecules [327]. Even in the daily life, SERS is already applied in the quality control of industrial processes [328], art preservation [329], and encoding luxury goods/bank notes as nanotags for brand/currency security [330].

When combined with microfluidic techniques, Raman spectroscopy is promising for biomolecular detection in a miniaturized lab-on-a-chip (LOC) device [331, 332]. Since thc analysis of small analyte volumes as well as the detection of low analyte concentrations is required in bioanalytics, the combination of SERS and a lab-on-a-chip device seems to be a promising approach to offer a sensitive detection with reproducible measurement conditions and a highly defined detection area specified by a chip system [333, 334, 335]. This Raman-on-chip concept will evoke the development of hand-held Raman spectrometers and disposable Raman systems, hence highlighting the commercial importance of Raman technique. In addition to the application of metallic nanoparticles as SERS condition in microfluidic devices [336, 337, 338, 339], the direct integration of SERS substrates in the LOC devices is of increasing interest [340, 341].

Many methods have been used to realize SERS conditions. A straightforward synthesis of gold and silver nanoparticles has been used extensively in aqueous solution. Raman enhancement factors of up to $\sim 10^{11}$, sufficient for a single-molecule detection, have been achieved in SERS "hot spot" (regions of the highest local electric field) associated with individual silver and gold colloids or in junctions between metal nanoparticles [234, 235]. However, this approach is limited by particle aggregation and suffers from the poor reproducibility between samples when dried onto an inert surface [342]. A way to overcome this limitation is to produce structured SERS substrate. So far, three main classes of SERS substrates have been developed: metallic rough surfaces [343, 344], assembled arrays of nanoparticles [345, 346, 347], and lithographically fabricated periodic nanostructures [333, 348]. Among various SERS substrates, a good compromise between moderate enhancement and high reproducibility can be offered by periodic nanostructures. Uniformly patterned periodic nanodome and nanopillar metallic/silicon structures have shown SERS enhancement factors of up to 10^8 on active substrates [348, 349, 350].

6.2 Laser-assisted replication of nanopillar arrays

Several techniques to fabricate periodic nanostructures are available. Electron beam lithography (EBL) offers maximum control over the nanostructures but suffers from low throughput and high costs. On the other hand, nanosphere lithography [351] does not provide complete control over the resulting nanostructure morphologies. Both methods cannot easily be integrated into the LOC fabrication. These problems can be solved through nanoimprint lithography (NIL) using a patterned silicon wafer as a reusable molding template. A subsequent gold coating is the only fabrication step requiring high vacuum. In this work, the gold-deposited cyclic olefin copolymer (COC) nanopillar arrays were fabricated by laser-assisted replication and subsequent gold film deposition. This method provides replication accuracy down to $\sim$ 20 nm and results in reproducible and efficient SERS nanostructures.

EBL was used for the fabrication of the silicon templates containing several sites with three patterns of inverted nanopillar (cylindrical nanowell) arrays. Poly(methyl methacrylate) (PMMA) 950k A2 (MicroChem) resist was spin-coated onto a 4-inch silicon wafer for 60 s at 1250 rpm and subsequently baked on a hotplate at 180 °C for 5 min. Cylindrical nanowell patterns for later nanopillar arrays were written onto the photoresist across several discrete areas of 0.5×0.5 mm^2 using an e-beam writer (Vistec Lithography Inc., Vistec VB6). In one 4-inch silicon wafer there were four regions of equal area size. Each region contains 7×9 individual nanowell array sites. Three different nanowell features were arranged on three of the four regions. The separation distance between adjacent nanowells was set at 20, 200 and $(D' + 20)$ nm, respectively, where D' is the nanowell diameter. The line and column separation distances between these sites were defined as 1.5 and 2 mm, as shown in Figures 6.1(a). Nine different nanow-

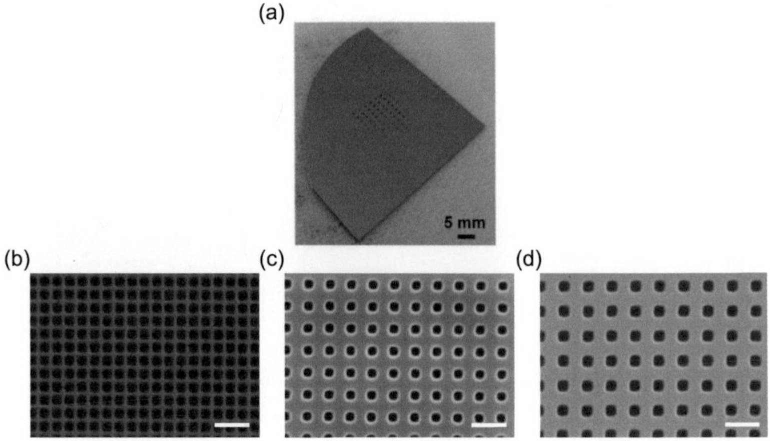

Figure 6.1: (a) Photograph of the fabricated silicon master stamp (after wafer dicing) with 7 × 9 nanopillar array fields. SEM images of a silicon master stamp with nanowells (inverted nanopillars) having a diameter D' of 150 nm and separation distances of (b) 20 nm, (c) 170 nm and (d) 200 nm. The scale bars correspond to 500 nm.

ell diameters were arranged in lines. And seven different e-beam doses were applied in columns (listed in Appendix A.3), in order to optimize the e-beam writing and following etching process. After development with methyl isobutyl ketone (MIBK) and isopropyl alcohol (IPA) (1:3) for 10 s, the resist structures were transferred into the silicon wafer by SF_6-based reactive ion etching. Nanowell arrays with a depth of $\sim$ 100 nm were finished after the PMMA resist was stripped. An anti-adhesion coating consisting of a self-assembled monolayer of fluorinated organosilanes completed the silicon template fabrication. The 4-inch silicon template was then divided by dicing into individual master stamps for laser-assisted replication. Figures 6.1(b)–(d) presents nanostructure details of one representative silicon master stamp.

A schematic illustration of the subsequent laser-assisted replication process is depicted in Figure 6.2. In comparison with common polymer sub-

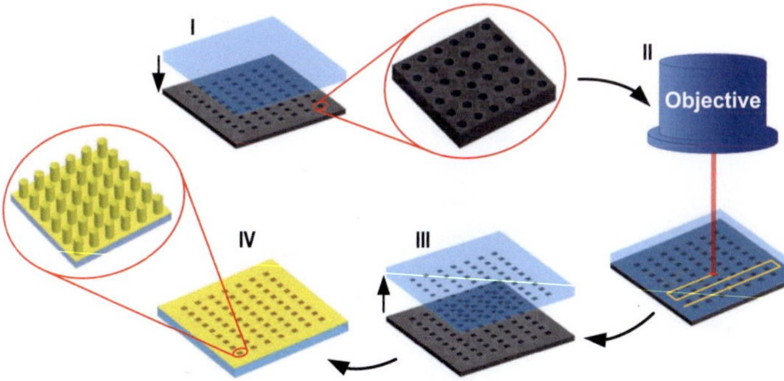

Figure 6.2: Schematic illustration of the laser-assisted replication process: (I) positioning of the COC substrate on a patterned silicon template, (II) laser-assisted replication, (III) detachment of the substrate and (IV) sputter-deposition of a thin gold-layer.

strate materials, like polydimethylsiloxane (PDMS) or poly(methyl methacrylate) (PMMA), COC exhibits relatively weak Raman signals, i.e., a weak interference with analyte signals [352, 353] and therefore it was chosen to build the SERS nanopillar arrays. A 1-mm-thick COC substrate with a size of 20×20 mm^2 was positioned on the template, covered with a glass plate (thus providing a contact pressure of 0.1 MPa) and exposed to the laser diode beam (50 W at 940 nm). The laser was focused to a ~ 0.3 mm^2 spot and scanned with a velocity up to 40 mm s^{-1}. Since the COC is transparent to the applied laser wavelength, the laser beam was mainly absorbed by the silicon surface in contact with the COC substrate. The laser scanning parameters were adjusted so that the local temperature of the irradiated polymer region was increased to exceed its glass transition temperature $T_g \approx 140$ °C. Above T_g, the nanostructures on the rigid template were (inversely) replicated into the polymer substrate. After cooling down, the polymer substrate with positively replicated nanostructures was detached from the silicon template and the process was repeated with a fresh COC substrate. A final gold sputter deposition completed the fabrication of a SERS-active substrate. Figure 6.3(a) presents a photograph of the completed SERS nanopillar array chip fabricated on a COC substrate comprising 7×9 nanopillar array sites. The contrast variance indicates different nanopillar patterns. The SEM image of one representative SERS nanopillar array site illustrates nearly defect-free replication, as shown in Figure 6.3(b). A stringent inspection of SEM images revealed that typically $\sim$ 2–10 % and rarely up to 30 % of all nanopillars were slightly bent, deformed or shifted. However, it is expected that these minor shape-defects have only little influence on the overall SERS performance of nanopillar arrays.

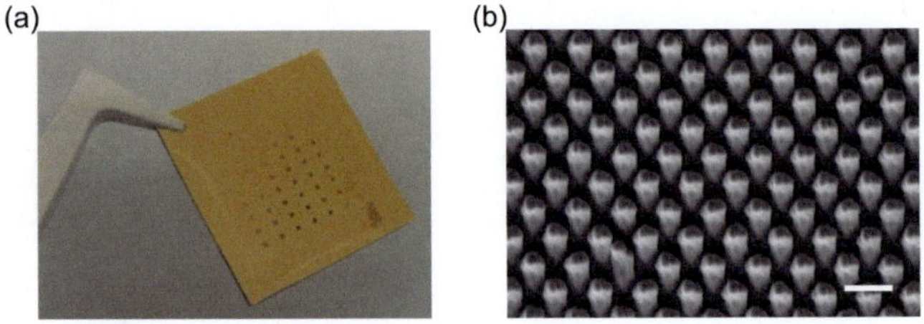

Figure 6.3: (a) Photograph of a gold-coated COC SERS substrate with 7×9 active nanopillar array sites. (b) SEM image of one active SERS site with nanopillars having a diameter of 200 nm and a spacing of 190 nm. The scale bar corresponds to 500 nm.

It is well known that the characteristic size of nanostructures and spacing between them in a periodic array can dramatically influence the SERS enhancement factors. In order to investigate the effect of both, different patterns of Au/COC nanopillars with diameter D, varying from 60 nm to 260 nm in steps of 20 nm, were fabricated. With regard to the feasible EBL resolution and replication accuracy, the minimum spacing between adjacent COC nanopillars was set at 20 nm. The typical thickness of deposited gold-films was measured to be about 25 nm on the top and 5 nm on the side walls of the nanopillars, respectively. Correspondingly, the minimum spacing between Au/COC pillars was $d \sim 10$ nm. In addition, arrays with larger interpillar spacing were fabricated, e.g., $d = 190$ nm and $d = D$, i.e., of the same value as the diameter D. Figure 6.4 shows SEM and atomic force microscope (AFM) images of three SERS nanopillar arrays with $D = 100$ nm and different interpillar spacings. The height of the COC nanopillars in all arrays was about 80 nm.

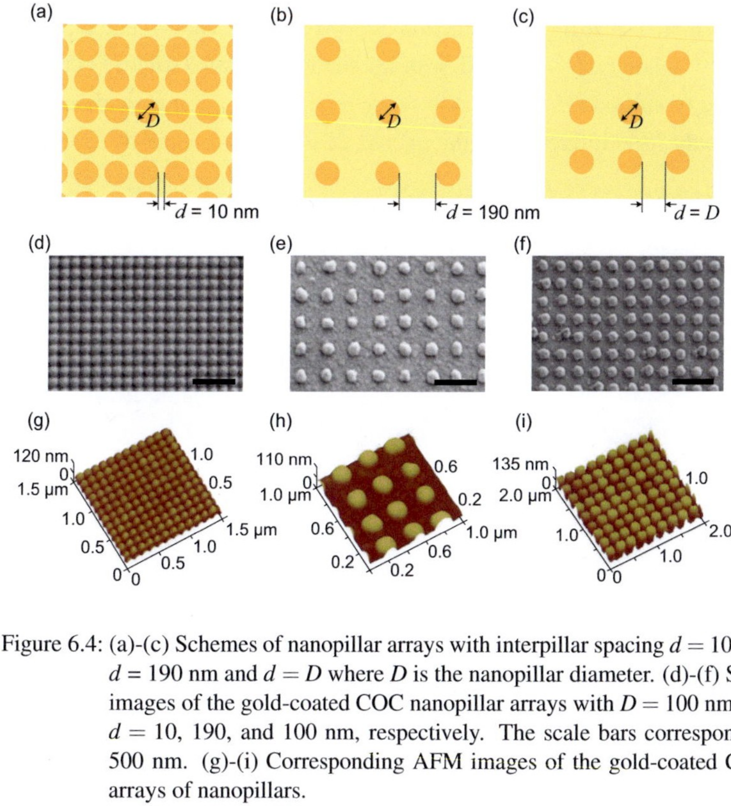

Figure 6.4: (a)-(c) Schemes of nanopillar arrays with interpillar spacing $d = 10$ nm, $d = 190$ nm and $d = D$ where D is the nanopillar diameter. (d)-(f) SEM images of the gold-coated COC nanopillar arrays with $D = 100$ nm and $d = 10$, 190, and 100 nm, respectively. The scale bars correspond to 500 nm. (g)-(i) Corresponding AFM images of the gold-coated COC arrays of nanopillars.

6.3 Computational analysis

As elucidated in Section 2.4.3, the main SERS enhancement mechanisms have been proposed to be of electromagnetic and chemical origin. In this case, the SERS effect arises mainly because the nanosized gap between nanopillars causes an intense localized electromagnetic field. From another perspective, a metallic nanopillar array can be considered as an array of optical antennas [354]. The induced *localized surface plasmon resonance* (LSPR) effect contributes to the enhancement both of the incident excitation and the Raman scattered light [221]. The local SERS electromagnetic enhancement factor (EF) scales approximately as [232]:

$$\mathrm{EF} \propto \frac{|E|^4}{|E_0|^4} \tag{6.1}$$

where $|E|$ is the amplitude of the enhanced local electric field and $|E_0|$ is the amplitude of incoming electromagnetic wave.

To elucidate the electromagnetic field distribution in the active SERS-substrates, the *finite-difference time-domain* (FDTD) solutions program (Lumerical, version 8.7.4) was used to simulate the near-field electric field distribution in modeled Au/COC nanopillar arrays excited with a normally incident plane wave at $\lambda = 632.8$ nm, polarized along the x-axis (as applied in the experiment). In this model, a 25-nm-thick gold-layer was assumed on top of the 80-nm-high COC nanopillars and a 5-nm-thick gold-layer on their side walls, as shown in Figure 6.5(a). To prevent numerical singularities in the form of electromagnetic "bursts" arising at the sharp edges and to take into account the realistic contours of the fabricated structures, the nanopillar edges were modeled by a curved surface with a radius of curvature of 3 nm. The boundaries of the simulation region in x, y-directions were set periodic to take the periodicity of the whole array into account. A 1 nm mesh in x, y, z-directions in the whole simulation region was found to be well suited for the convergence of the calculations. The cross-sectional electrical intensity profiles were obtained by a XZ-monitor area through the nanopillar axis.

The simulation results for nanopillars with a diameter $D = 100$ nm and three different interpillar spacings are shown in Figure 6.5(b)–(d) with the color-coded intensity representing the normalized amplitude of the en-

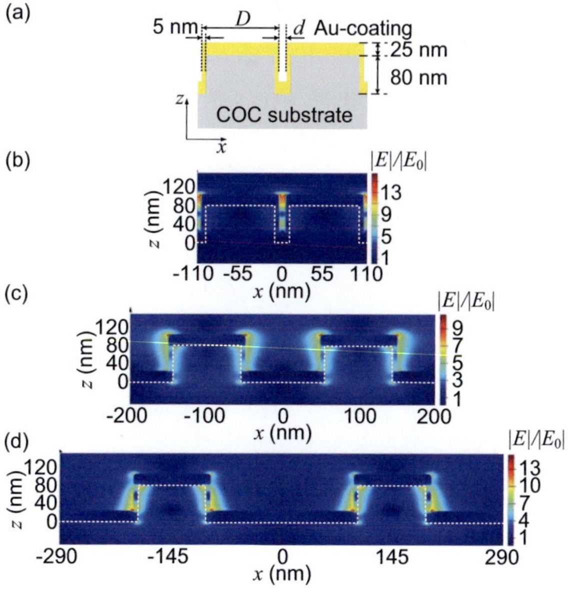

Figure 6.5: (a) Schematic side view of Au/COC nanopillars model for FDTD simulations. (b)-(d) Color-coded FDTD-simulated images of the near-field electric field intensity distributions for pairs of Au/COC nanopillars with diameter $D = 100$ nm and spacing $d = 10$, 100, and 190 nm, respectively. The dashed lines denote the COC nanopillar boundaries. $z = 0$ corresponds to the bottom of uncoated COC nanopillars.

hanced electric field $|E|$ with respect to the amplitude of the incoming electric field $|E_0|$. For nanopillars with spacing of 10 nm, the enhanced electric field regions with two antinodes are clearly seen within the gaps between adjacent nanopillars (see Figure 6.5(b)). This minimum spacing also offered the highest electric field enhancement, as could be expected due to the surface plasmon coupling effect [232]. When the spacing exceeded ~ 30 nm, the major electric field enhancement could only be found in the vicinity of nanopillar side walls, indicating a weakened electromagnetic coupling between nanopillars, as shown in Figure 6.5(c) and 6.5(d).

Furthermore, enhanced field regions were also found to be generated in-side nanopillars and thus "inaccessible" for SERS of analyte molecules. When the nanopillar diameter exceeded ~180 nm and the interpillar spacing was similarly large, the enhanced electric field regions were mainly confined within nanopillars, as indicated in Figure 6.6 and practically did not contribute to the SERS effect. These trends are reflected in Figure 6.7,

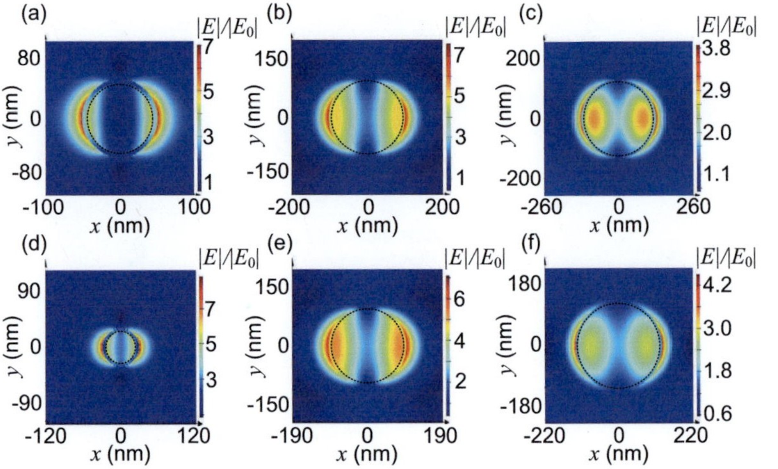

Figure 6.6: FDTD-simulated images of near-field electric intensity distributions around Au/COC nanopillars in the XY-plane. For arrays with the same gap distance as the nanopillar diameter, D: (a) $D = 100$ nm, (b) $D = 200$ nm and (c) $D = 260$ nm. With the gap distance of 190 nm: (d) $D = 60$ nm, (e) $D = 200$ nm and (f) $D = 260$ nm. The dashed lines denote the COC nanopillar boundaries. The excitation wavelength is 632.8 nm.

which shows the average electric field enhancement in "hot spot" regions as a function of nanopillar diameter varying from 60 to 260 nm for three groups of arrays having constant spacings of 10 and 190 nm, respectively, and those with the spacing following the diameter ($d = D$). The "hot spot" region was defined as an area, where the enhanced electromagnetic field in the exterior of Au/COC nanopillars was more than $1/e$ of the maximum value. Such a definition of the average electric field enhancement appears to provide a more realistic estimation of the SERS effect than the maximum value. Another important factor determining the overall SERS efficiency is the relative "hot spot" area with regard to the whole SERS array area, as discussed in detail in Section 6.4.

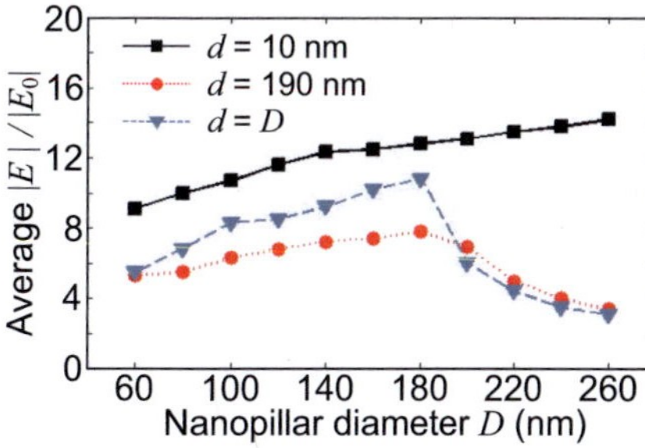

Figure 6.7: FDTD-simulated electromagnetic field enhancement averaged over "hot spot" regions versus nanopillar diameter for three different nanopillar array patterns.

6.4 SERS experiments with nanopillar arrays

The SERS effect of various Au/COC nanopillar arrays was first experimentally characterized using the laser dye rhodamine 6G (Rh6G) as the analyte and a helium-neon (He-Ne) laser with an emission wavelength of 632.8 nm as the excitation source. The measurements were performed on the modular micro-Raman system (WITec, CRM200) usually with a He-Ne excitation laser coupled into the microscope via a single-mode optical fiber. Typically, a 20× objective lens (NA = 0.40) was applied, resulting in a laser spot diameter of $\sim 4\,\mu m$ on the sample. In this apparatus, the He-Ne laser could be readily exchanged by an organic semiconductor laser emitting at the nearby wavelength of 631.4 nm as described in detail below. SERS samples were prepared directly before measurements by drop-casting and uniformly spreading 30 µl of a 10 µM aqueous solution of Rh6G over the whole ($20 \times 20\ mm^2$) nanopillar array SERS substrate. The substrate was then dried under ambient conditions. The average density of the deposited Rh6G molecules corresponded to ~ 0.45 molecules nm^{-2}.

Figure 6.8 show SERS spectra of Rh6G molecules, which were acquired for different nanopillar arrays under the same experimental conditions. In accordance with the FDTD simulation results, the SERS signals of Rh6G exhibited a significant dependence on the nanopillar array geometry. In case of the arrays with the minimum interpillar spacing of $d = 10$ nm, the SERS intensity first increased with the increasing nanopillar diameter up to $D \sim 180$ nm, but then started to decrease for the larger diameter values, as seen in Figure 6.8(a) and 6.9(a). For a set of arrays having nanopillar spacing equal to their diameter ($d = D$), the SERS intensity monotonically decreased with increasing diameter from 60 to 220 nm, as seen in Figure 6.8(b) and 6.9(b). For comparison, similar measurements were performed with commercially available Klarite® SERS substrates (Renishaw Diagnostics). It was found that the optimum nanopillar array patterns (i.e., $d = 10$ nm, $D \sim 180$ nm) provided SERS signals which were stronger by a factor of ~ 20 than those from Klarite® SERS substrates.

For evaluation of the absolute SERS enhancement factors (EFs), the SERS response to the "standard" Raman signals of the analyte (Rh6G) were compared - taking into account the respective numbers of Rh6G molecules probed within the laser focus. In this comparison, the SERS signals are assumed to only arise from molecules occupying the calculated "hot spot" regions of a given SERS substrate. The SERS enhancement factor (EF) of Rh6G on Au/COC nanopillar arrays was estimated for the SERS peaks at 614 cm^{-1} and 1367 cm^{-1} (shifted to 620 cm^{-1} and 1381 cm^{-1}, respectively, for Rh6G dried on a gold-coated unstructured COC substrate) using Eq. 2.47. The laser excitation/detection volume in the reference solution, which was estimated as $\sim 200\,\mu m^3$ for the microscope setup with an objective (20×, NA = 0.4). 30 µl of a 10 mM Rh6G aqueous solution was applied to a gold-coated unstructured COC substrate (reference sample). The total number of Rh6G molecules in this solution was about 1.8×10^{17}. For an excitation/detection volume of $\sim 200\,\mu m^3$, N_{vol} was calculated as

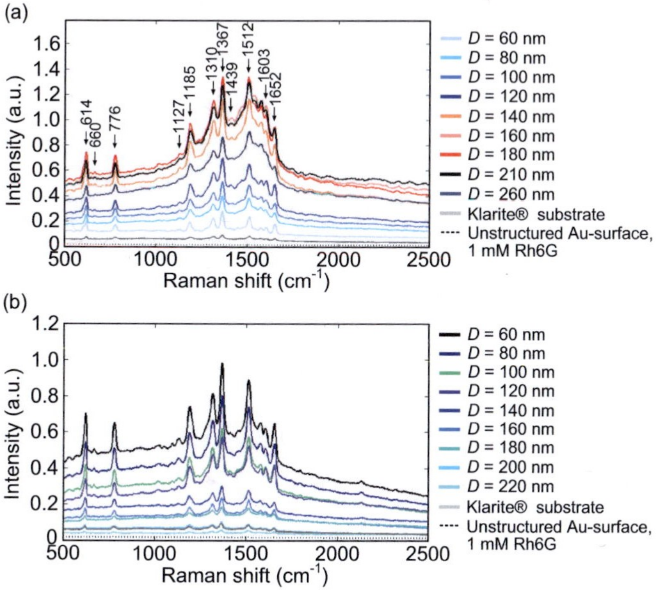

Figure 6.8: SERS spectra of Rh6G deposited on arrays with different nanopillar diameters *D*, and with interpillar spacings: (a) $d = 10$ nm and (b) *d* equals to *D*. The spectra are compared with those of Rh6G similarly deposited on a commercial Klarite® SERS substrate and of a solid Rh6G layer deposited from a 1 mM solution onto an unstructured gold surface. All spectra were measured under the same conditions: 4 mW excitation at 632.8 nm, 10 s acquisition time.

$\sim 1.2 \times 10^9$. Average signals $I_{\mathrm{ref}}(620\ \mathrm{cm}^{-1})$ and $I_{\mathrm{ref}}(1381\ \mathrm{cm}^{-1})$ were measured as 2290 and 4030 counts, respectively (5 mW excitation at 632.8 nm, 10 min acquisition time).

N_{surf} was calculated as the number of Rh6G molecules deposited in the "hot spot" regions of Au/COC nanopillars within the laser spot. These regions were defined from modeling as described above. For instance, for the Au/COC nanopillars with $d = 10$ nm and $D = 260$ nm, the "hot spot" area amounted to ~ 2000 nm^2 per nanopillar. Experimentally, a 10 nM diluted solution of Rh6G was applied and dried over the SERS

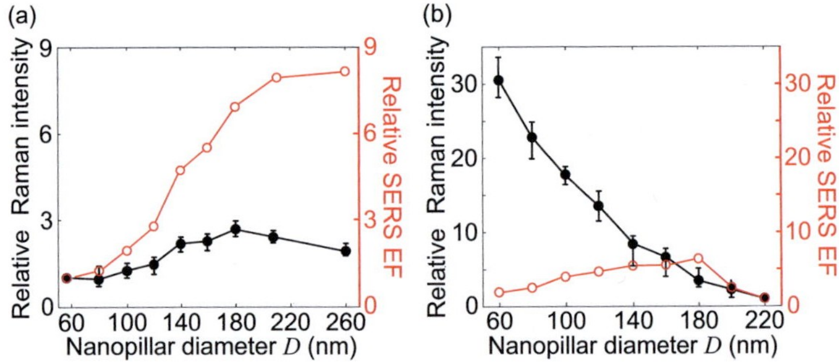

Figure 6.9: Relative SERS signals and corresponding calculated “hot spot” average enhancement factors (EFs) as a function of nanopillar geometries for arrays with: (a) $d = 10$ nm and (b) $d = D$, respectively.

substrate (20 × 20 mm^2). With regard to the very low analyte concentration (10 nM) applied, a roughly homogeneous, non-site-specific analyte deposition/adsorption of the Rh6G molecules on the whole SERS substrate with Au/COC nanopillar arrays was assumed. Since the SERS areas (7 × 9 sites, each with a size of 0.5 × 0.5 mm^2) occupy only a small part of the whole substrate (20 × 20 mm^2), their contribution to the whole surface area is negligible. As a result, we approximate that the adsorbate density (per surface unit) is nearly the same for nanoarrays with different geometries. By depositing a 30 µl volume of a 10 nM diluted solution, this corresponded to about 4.5×10^{-4} molecules/nm^2. Within one laser excitation spot (diameter of ~ 4 µm), there are about 150 nanopillars with $d = 10$ nm and $D = 260$ nm. The number of molecules within the “hot spot” regions N_{surf} is about 140. The SERS measurement results showed $I_{SERS}(614\ cm^{-1}) = 2990$ counts and $I_{SERS}(1367\ cm^{-1}) = 4670$ counts (5 mW excitation at 632.8 nm, 10 min acquisition time). Hence, the SERS enhancement factors are EF(614 cm^{-1}) $\sim 1 \times 10^7$ and EF(1367 cm^{-1}) $\sim 9 \times 10^6$. Note that this result is about a factor of ~ 200 higher than the EF calculated according to Eq. 6.1 as the fourth power of the modeled electric field enhancement averaged over “hot spot” regions (average $|E|/|E_0| \sim 15$, as shown in Figure 6.7). Such a discrepancy between the measured and calculated SERS EFs has often been seen and is usually attributed to the additional contribution of chemical effects which are not considered in the simulation [232]. One can also expect that the gold surface on the nanopillars is actually not as smooth as that of the model structures. Such surface roughness would lead to higher local electric field enhancements than predicted.

For nanopillar arrays with $d = 10$ nm, Figure 6.9(a) compares the experimental dependence of the SERS intensity (peak at 614 cm^{-1}) on array geometry (nanopillar diameter) with the calculated SERS "hot spot" average enhancement factors. The experimental intensity and calculated EFs are normalized to the values for the nanopillar array $d = 10$ nm, $D = 60$ nm and are shown in relative units. Both curves demonstrate a roughly similar trend. However, the measured SERS intensity increases more slowly than the calculated "hot spot" EFs and even slightly decreases for $D \geq 180$ nm. This discrepancy can be explained by the fact that, according to FDTD simulations, the relative area of the "hot spots" (versus the whole array area) decreases upon increasing D. Correspondingly, fewer molecules would on average occupy the "hot spot" regions and thus contribute to the SERS signal (under the assumption of approximately uniform, non-site-specific analyte deposition/adsorption onto the whole nanopillar array surface). According to simulations, the relative "hot spot" area is reduced by a factor of 4.3 by increasing D from 60 to 260 nm. Above $D \sim 180$ nm, the decrease of the "hot spot" area overtakes the increase of EF and the overall SERS efficiency starts to decrease.

The effect of the relative area of the "hot spot" regions can be further verified for the group of nanopillar arrays having equal nanopillar diameter and spacing ($d = D$), as shown in Figure 6.9(b). The experimental results and calculated EFs are normalized to the values for the nanopillar array with $d = D = 220$ nm. The modeled EF in the "hot spots" rises relatively slowly when the nanopillar diameter increases in the range of $D \sim 60 - 180$ nm. Beyond $D \sim 180$ nm, the major enhancement of the electromagnetic field occurs only within nanopillars and the weakened coupling between nearby nanopillars results in a decrease of "hot spot" EFs. In parallel, the relative "hot spot" area diminishes strongly, even faster than in the case of d = 10 nm as discussed above. Hence, the total SERS efficiency drops with increasing D. The occupancy or relative area of the "hot spots" should not be neglected as it can play a significant role in determining the overall efficiency of a SERS substrate.

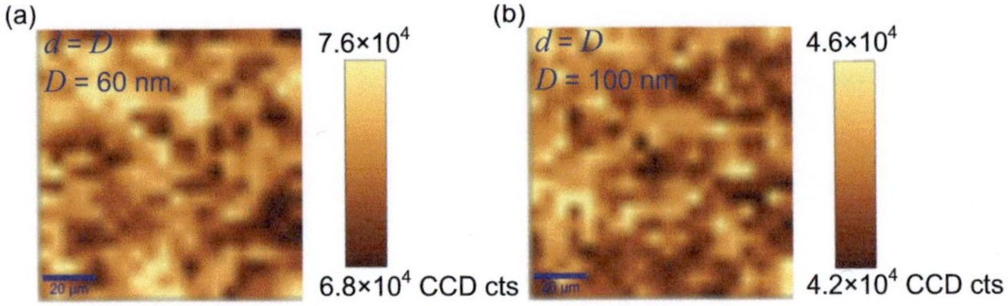

Figure 6.10: Typical SERS images of arrays with equal nanopillar spacing and diameter ($d = D$): (e) 60 nm and (f) 100 nm, respectively. The color-coded intensity corresponds to the area of the 614 cm^{-1} peak of Rh6G. Lateral resolution and pixel size are $\sim 2\,\mu m$ and $2 \times 2\,\mu m^2$, respectively.

To characterize the lateral variation of the SERS response across nanopillar arrays, i.e., the uniformity of the SERS substrates over length scales comparable to the laser focus, Raman images were acquired by XY-scanning the samples under a microscope objective with the aid of a piezo-stage. The scanned area and lateral resolution were 0.1×0.1 mm^2 and $\sim 2\,\mu m$ (20$\times$ objective), respectively. Figure 6.10 shows the color-coded images of the characteristic Raman peak of Rh6G at 614 cm^{-1} after subtraction of the broad background signal. Raman images acquired for several nanopillar arrays exhibited intensity variations in the SERS signals of less than 10 %. This observation provides further evidence of the high accuracy and reproducibility of the employed laser-assisted replication process.

6.5 Organic semiconductor laser SERS experiments

The SERS experiments with DFB-OSLs were achieved in a modular micro-Raman system (WITec, CRM200), the same setup as for using the He-Ne laser as the excitation source. The exchange from the He-Ne laser to an organic laser emitting at 631.4 nm can be easily performed via changing the integrated module comprising an appropriate clean-up bandpass filter. The organic laser module was introduced in Section 5.4.1. The overall fiber coupling efficiency was as high as ~ 40 %. The excitation power on the analytes could reach up to 250 µW.

In contrast to a standard He-Ne laser, the OSL is a quasi-continuous-wave Raman excitation source which emits short pulses ($\sim$ 1 ns) at a repetition rate of 10 kHz as determined by the DPSS UV-pump laser (Crylas, FTSS355-Q2). In addition, to obtain a high fiber coupling efficiency, a multi-mode fiber with a core diameter of 50 µm was used in the excitation path, which generates an enlarged excitation spot on the sample in the setup ($\sim$ 20 µm with a 20$\times$ objective) as compared to the He-Ne laser (coupled via a single-mode fiber with core diameter of 3 µm). The image of this spot as produced by a tube lens in the Raman microscope was also correspondingly enlarged. However, the influence of excitation spot size was proven insignificant for SERS detection. As illustrated in Figure 6.11, similar SERS signals of Rh6G were detected using both lasers at the same excitation power.

Finally, the organic semiconductor DFB laser was used for the biomolecular SERS detection of adenosine deposited on one of the most efficient nanopillar arrays (d = 10 nm and D = 100 nm). Figure 6.12 shows the Raman spectra of adenosine molecules deposited by drop-casting from different aqueous solutions with concentrations ranging from 1 µM to 10 mM (using 30 µl volumes in each case). The detection limit for dried adenosine deposits prepared in this fashion was $<$ 300 pmol - at the excitation power of 0.2 mW and a relatively short acquisition time of 10 s.

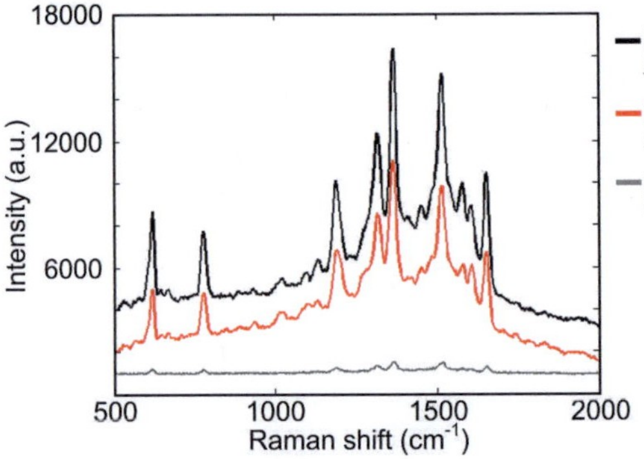

Figure 6.11: SERS spectra of Rh6G (drop-cast from 30 μl of a 10 μM aqueous solution) as measured on Au/COC nanopillar arrays (d = 10 nm, D = 100 nm), upon excitation with an organic semiconductor DFB laser and a He-Ne laser (both at 0.2 mW), in comparison with the Raman spectra of Rh6G on a Klarite® SERS substrate (also drop-cast from 30 μl of a 10 μM aqueous solution). Measurement conditions: 20× objective, 600 g mm^{-1} grating, 5 s acquisition time.

6.6 Discussion

The plasmonic characteristics of nanopillar arrays can be investigated via extinction and reflectivity measurements [350, 355]. However, it is not easily possible to measure the extinction spectra of these samples. As an opaque gold layer was sputtered on the whole sample, transmission measurements are in general not possible. The optical characterization in reflection geometry, is also very demanding, as the active sample area is only (0.5 × 0.5 mm^2). This precludes the use of a conventional spectrometer and would require a microscopic apparatus suitable for such measurements. In this work, the surface plasmon resonance behavior of the Au/COC nanopillars was first characterized by measuring their scattering properties with white-light illumination dark-field microscopy. This method was applied previously at LTI for investigating the resonance properties of gold dipole nanoantennas [356]. The characterization setup is presented in Figure 6.13(a). A microscopy halogen lamp (Zeiss HAL 100) homogeneously

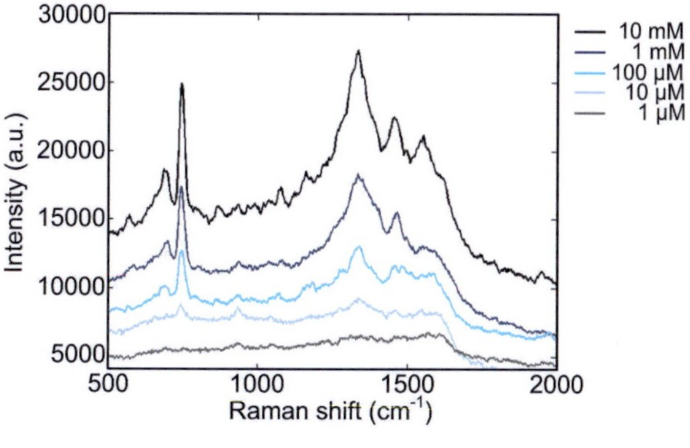

Figure 6.12: SERS spectra of adenosine deposited on Au/COC nanopillar arrays (d = 10 nm, D = 100 nm) from aqueous solutions with concentrations ranging from 1 μM to 10 mM. SERS was excited with an organic semiconductor DFB laser at 631.4 nm at a power of 0.2 mW. Other measurement conditions: 20× objective lens, 600 g mm^{-1} grating and 10 s acquisition time.

illuminated a dark-field condenser (Zeiss, EC Epiplan NEOFLUAR, 50×, NA = 0.8). The dark-field scattering from the nanostructures was then collected by the same objective and reflected onto an electron-multiplying CCD (EMCCD) camera (Andor iXon) using an imaging spectrometer (Acton 2500i). The slit opening of the spectrometer was then used to selectively image lines of nanopillar arrays onto the EMCCD detector. The measurement results in Figure 6.13(b) indicate a strong resonant enhancement at wavelength around 650 nm. For a spacing of $d = 10$ nm, the nanopillar resonance is found to be dependent on the nanopillar diameter D. The scattering peaks shift towards longer wavelengths for larger nanostructures and the peak intensities increase for larger nanostructures.

Additionally, FDTD-simulations were performed to further check the wavelength-dependent plasmonic effect on the electric field intensities. For this purpose a two-dimensional monitor (20 nm × 80 nm) normal to the x-axis was placed between two neighbor nanopillars. The simulation results also predict a significant dependence of the optimum excitation wavelength

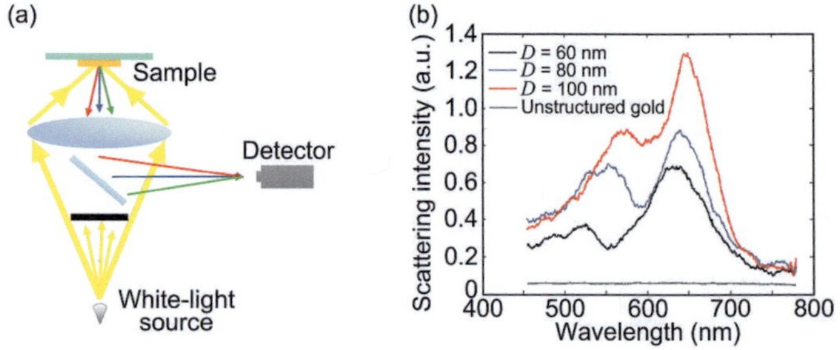

Figure 6.13: (a) Dark-field microscopy in reflection configuration. The white-light continuum (generated with a halogen lamp) used for excitation is shown with yellow arrows, while the scattered light generated at the nanostructures of the sample is depicted in red, green, and blue. (b) Scattering spectra of the Au/COC nanopillar arrays with interspacing of 10 nm and different diameters.

on the nanoarray geometry, as shown in Figure 6.14. For instance, optimum excitation at $\sim$ 600–650 nm (i.e., close to the laser wavelength applied in this work) was revealed for the nanopillars with $d = 10$ nm and $D = 20$ nm. Increasing the nanopillar diameter D, while keeping the same spacing of $d = 10$ nm results in the red shift of the excitation resonance.

The dark-field scattering characterization and FDTD simulations predict that the spectral resonance peaks of the nanopillar arrays are slightly higher than the current laser wavelength. Consequently, the excitation at longer wavelengths ($\sim$ 750 nm and above) might be even more favorable. These nanoarrays may be of significant interest for near-infrared (NIR) SERS analyses. It is noted that the NIR excitation range is currently practically inaccessible for organic semiconductor lasers (efficient lasing till $\sim$ 700 nm). However, as the dense nanopillar arrays allow the excitation of several modes in the visible spectral range, it is promising to employ organic tunable lasers as excitation sources to study the excitation wavelength dependence in future.

In this chapter, organic semiconductor DFB lasers were applied in SERS-based (bio)analytics. The SERS condition was achieved by substrates con-

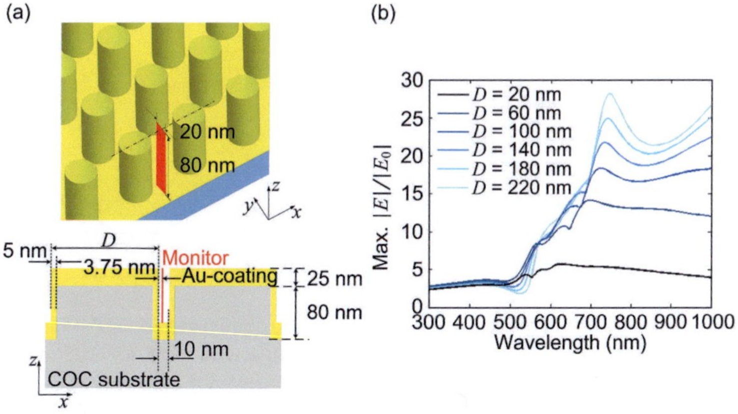

Figure 6.14: (a) Schematic views of the Au/COC nanopillars model for FDTD simulations. An x-normal 2D-monitor (20 nm $\times$ 80 nm) was used check the wavelength-dependent electric fields. (b) FDTD-simulated maximum electromagnetic field enhancement over the two-dimensional monitor on 25 nm gold-film coated COC nanopillars with spacing of 10 nm.

sisting of arrays of gold coated cyclic olefin copolymer (COC) nanopillars with variable diameters ranging from 60 nm to 260 nm. Fabricated by laser-assisted replication, the spacing between adjacent nanopillars was precisely controlled down to $\sim$ 10 nm. The employed silicon master templates were prepared by electron beam lithography technique and could be applied multiple times in the replication process. The replication is fast and potentially low-cost. This fabrication method may contribute to the advancement of analytical SERS applications. Furthermore, such polymer SERS substrates may be directly integrated into, e.g., microfluidic channels of the photonic devices. SERS enhancement factors estimated to be up to $\sim 10^7$ for Rh6G were determined for the "hot spot" regions in the nanopillar arrays-based on FDTD modeling of their lateral extent. The modeling also supported the experimentally characterized dependence of the SERS enhancement on the nanopillar array geometry (nanopillar diameter and spacing). As a first application of this approach, the sensitive concentration detection of adenosine using an organic semiconductor DFB laser as the excitation source was demonstrated. This combination of metallic/polymer surface-enhanced Raman nanopillars and organic excitation lasers appears to be particularly promising for routine SERS biomolecular analytic in future miniaturized lab-on-a-chip devices.

7 Summary and Outlook

7.1 Summary

In this thesis, organic semiconductor lasers (OSLs) have been studied with special emphasis on the exploration of new application fields in Raman spectroscopy. It has been shown, that optically pumped organic lasers exhibit a low laser threshold and a single longitudinal emission mode with narrow spectral bandwidth. These features are determined by well-defined distributed feedback resonators. The laser emission stability and the average laser output power can be improved by variation of organic media, resonator features and optical pumping conditions. Parallel, time-saving fabrication approaches have been developed to create miniaturized spatially defined laser pixels for lab-on-a-chip spectroscopy applications. The results of this work suggest further investigations concerning spectroscopic applications as well as basic research related to laser lifetime and laser output power of organic semiconductor lasers.

The main results are summarized as follows:

- Two technologies, ink-jet printing for the local definition of active materials and laser-assisted replication for the laterally controlled fabrication of DFB-corrugations were presented in the fabrication of spatially defined organic semiconductor lasers. As the first demonstration of ink-jet-printed organic semiconductor lasers, a blend of high-boiling and low-boiling solvents was used for dissolving the polymer to optimize the printing uniformity. The ink-jet-printed laser devices showed laser thresholds as low as 76 nJ pulse^{-1} and a homogeneous laser emission with a standard deviation of 0.5 nm over large areas. The laser emission properties were comparable to those of organic semiconductor lasers fabricated via established solution-processing methods. Additionally, ink-jet printing allows the deposition with arbitrary lateral shapes and small lateral dimensions. Thus an accurate deposition of the organic active medium on grating trenches with no contamination of other parts of the functional chip could be achieved. In the second approach, a novel nanoimprint lithography method, laser-assisted replication, was employed for the fabrication of spatially defined DFB-corrugations. With the help of an infrared laser beam instead of large-area heating, it allows a spatially defined pixel-sized replication and a short processing period of 5 min for a replication area of 30×30 mm^2. Using this technique, surface-emitting laser pixels based on second-order DFB nanogratings were accomplished with different periods in lateral dimension of 0.5×0.5 mm^2 on a PMMA substrate. Novel coupled edge-emitting organic lasers were achieved by evaporating Alq_3:DCM on both of first-order DFB-corrugations and neighboring microscale waveguide trenches fabricated through laser-assisted replication. In comparison with the conventional deep ultraviolet induced PMMA waveguides, the waveguides from the same organic laser material are advantageous for an optimum light coupling efficiency. It was also proven that using this fabrication method, further nanostructures could be added into the chip platform without negative influence on neighboring photonic components.

- For applications in Raman spectroscopy, Alq_3:DCM organic lasers were built on qualified one-dimensional glass grating with a period of 400 nm. The DFB-OSLs showed a laser slope efficiency as high as 7.6 % and offered an excitation power up to 350 μW. In the free-beam coupling configuration, the bottom-emitting organic laser was directly guided into an inverted microscope to excite the Raman spectra of the analytes. In the fiber coupling configuration, the bottom laser emission was additionally harvested by adding a reflector film on the bottom of the DFB grating. It was together with the top surface laser mission coupled into a fiber-connected upright optical microscope. Using these two free-space setup configurations, the DFB-OSLs were successfully applied as excitation sources in Raman spectra analysis of sulfur, cadmium sulfide and cyclohexane. One organic semiconductor tunable laser was introduced to facilitate a specific laser wavelength to match the commercially available optical filters designed for use with He-Ne lasers at 632.8 nm.

- In the case of Raman detection of biomolecular analytes, the surface-enhanced Raman scattering (SERS) effect was applied to overcome the challenge of typically low analyte concentrations and improve the detection sensitivity. To achieve an efficient SERS condition, a novel gold-deposited polymer substrate comprising various nanopillar arrays were fabricated via laser-assisted replication. The nanopillar diameters were variably ranged from 60 nm to 260 nm and the spacing between adjacent nanopillars was precisely controlled down to ~ 10 nm. Based on the FDTD modeling of the electromagnetic fields and the experimental characterizations, SERS enhancement factors on rhodamine 6G Raman peaks were determined to be up to $\sim 10^7$ in the "hot spot" regions of the nanopillar arrays. The modeling also indicates a strong dependence of SERS enhancement on the nanopillar array geometry. As the first application of this approach, a sensitive concentration detection of adenosine down to < 300 pmol was demonstrated by using Au/COC nanopillar arrays as SERS substrate and an organic semiconductor DFB laser as excitation source.

7.2 Outlook

This work paves the way to apply organic semiconductor DFB lasers in Raman spectroscopy and in further LOC spectroscopic applications. More interesting topics are foreseen in both, basic research and technical processing of organic semiconductor lasers.

For spectroscopic applications, especially for Raman detection, the limited average laser output power is the main issue of organic semiconductor lasers. There are two feasible solutions available to enhance the emission power. One is to further increase the average output power by increasing the optical pump pulse energy or accelerating the repetition rate. However, in consideration of laser lifetime and emission stability, this method is probably not the appropriate one for a long-term reliable system response. Another idea is to integrate the pulsed organic lasers in a time-gated system for Raman spectroscopy. In the pulsed-laser Raman spectroscopy, a pulsed laser is used to excite the sample instead of a cw laser and the signal detection system is turned on only during the time intervals occupied by the laser pulses. During all other times the detection system is inactive. In effect, all of the Raman-information-bearing signal is compressed into narrow time intervals [357]. This concept has been widely used in Raman detections to mainly eliminate the strong fluorescence background of the analytes excited by femtosecond or picosecond laser pulses. Nowadays novel time gating systems have emerged to address the advantages of this technique, e.g., the ultra-fast Kerr shutter [358, 359] and the CMOS single-photon avalanche detector (SPAD) [360, 361]. The moderate pulse power of organic semiconductor lasers is below the photodamage threshold for biological living samples. Applying a gentle femto- or picosecond pulsed excitation with organic lasers is also favorable for stabilizing the emission output and prolonging the laser lifetime. In summary, it is prospective to apply organic semiconductor lasers in pulsed-laser Raman spectroscopy in the next stage.

The wide tunability in the visible spectral range is the main advantage of organic semiconductor lasers. This indicates further potential application fields of organic lasers in resonance Raman spectroscopy (RRS) and surface-enhanced resonance Raman spectroscopy (SERRS). For such applications, a wide-range tunable notch-filter is required for blocking the laser refection and Rayleigh scattering from the Raman signals. There have been already several designs of tunable notch filters in the research lab, e.g., the holographic notch filters [362, 363] and filters based on volume Bragg gratings (VBG) [364, 365]. More achievements on this issue will facilitate the application of tunable organic semiconductor lasers in RRS. Meanwhile, as discussed in this thesis, a basic research on device performance of tunable organic lasers is required for spectroscopic applications. Until now, numerous approaches of tunable organic lasers have been reported. However, a deep understanding and observation of variation of laser threshold, spectra width, lifetime, etc. along with the tuning procedure is still missing. This calls for a comprehensive study on this topic.

So far there have been no integrated laser excitation sources realized in an on-chip Raman platform. Organic semiconductor lasers are very good candidates for such purpose. Research should be addressed on the suppression of strong fluorescence from the organic active media. In addition to the perturbing signal from the laser itself, the all-polymer lab-on-a-chip platform delivers unavoidably its own Raman signals, i.e., interference with the analyte Raman signals. In comparison to the free-space configuration, it is an issue to add standard optical lens and notch filters into an on-chip Raman system. The developments, not only on integrated organic lasers, but also on other integrated optical passive elements should be key to the final solution. As elucidated in this thesis, the lasers based on polymer grating substrates suffer from the problems of photodegradation. The performance of integrated on-chip organic lasers should be further improved as qualified excitation sources for spectroscopic applications.

The SERS approach is a good tool to improve the Raman detection sensitivity. It is promising to evolve the SERS-based optofluidic platforms incorporating the fabricated nanopillar arrays. Utilizing laser-assisted replication, the microfluidic channels/networks and the nanostructure arrays can be simultaneously created on a polymer chip, allowing a mass production of SERS-active sites. Such microfluidic systems containing immobilized nanostructures are more stable in a biofluid analysis than colloid-based systems due to variations in the ionic strength of the biofluid [366]. The nanopillar arrays fabricated by laser-assisted replication are compatible with the lab-on-a-chip concept and offer the possibility of high-throughput analysis.

In conclusion, many advantages and many opportunities for organic lasers are foreseen in Raman spectroscopy and on-chip spectroscopic applications.

Bibliography

[1] T. H. Maiman, "Stimulated optical radiation in ruby," *Nature*, vol. 187, pp. 493–494, 1960.

[2] W. Demtröder, *Laser spectroscopy: basic concepts and instrumentation*. Berlin: Springer, 2003.

[3] C. Raman and K. Krishnan, "A new type of secondary radiation," *Nature*, vol. 121, pp. 501–502, 1928.

[4] G. Landsberg and L. Mandelstam, "Eine neue Erscheinung bei der Lichtzerstreuung in Krystallen," *Naturwissenschaften*, vol. 16, pp. 557–558, 1928.

[5] I. R. Lewis, Ed., *Handbook of Raman spectroscopy : from the research laboratory to the process line*. New York: Marcel Dekker, 2001.

[6] P. Sorokin and J. Lankard, "Stimulated emission observed from an organic dye, chloro-aluminum phthalocyanine," *IBM Journal of Research and Development*, vol. 10, pp. 162–163, 1966.

[7] F. P. Schäfer, W. Schmidt, and J. Volze, "Organic dye solution laser," *Applied Physics Letters*, vol. 9, pp. 306–309, 1966.

[8] B. H. Soffer and B. B. Mcfarlan, "Continuously tunable narrow-band organic dye lasers," *Applied Physics Letters*, vol. 10, p. 266, 1967.

[9] N. Karl, "Laser emission from an organic molecular crystal," *Physica Status Solidi (a)*, vol. 13, pp. 651–655, 1972.

[10] O. S. Avanesjan, V. A. Benderskii, V. K. Brikenstein, V. L. Broude, L. I. Korshunov, A. G. Lavrushko, and I. I. Tartakovskii, "Anthracene-crystals under intensive optical-pumping," *Molecular Crystals and Liquid Crystals*, vol. 29, pp. 165–174, 1974.

[11] C. K. Chiang, C. R. Fincher, Y. W. Park, A. J. Heeger, H. Shirakawa, E. J. Louis, S. C. Gau, and A. G. MacDiarmid, "Electrical conductivity in doped polyacetylene," *Physical Review Letters*, vol. 39, pp. 1098–1101, 1977.

[12] N. Tessler, G. J. Denton, and R. H. Friend, "Lasing from conjugated-polymer microcavities," *Nature*, vol. 382, pp. 695–697, 1996.

[13] F. Hide, M. A. Díaz-García, B. J. Schwartz, M. R. Andersson, Q. Pei, and A. J. Heeger, "Semiconducting polymers: A new class of solid-state laser materials," *Science*, vol. 273, pp. 1833–1836, 1996.

[14] S. V. Frolov, O. Masanori, G. Werner, Z. V. Vardeny, and Y. Katsumi, "Mirrorless lasing in conducting polymer poly(2,5-dioctyloxy-p-phenylenevinylene) films," *Japanese Journal of Applied Physics*, vol. 35, pp. L1371–L1373, 1996.

[15] V. G. Kozlov, V. Bulovic, P. E. Burrows, and S. R. Forrest, "Laser action in organic semiconductor waveguide and double-heterostructure devices," *Nature*, vol. 389, pp. 362–364, 1997.

[16] M. Berggren, A. Dodabalapur, R. E. Slusher, and Z. Bao, "Light amplification in organic thin films using cascade energy transfer," *Nature*, vol. 389, pp. 466–469, 1997.

[17] M. B. Christiansen, M. Schøler, and A. Kristensen, "Integration of active and passive polymer optics," *Optics Express*, vol. 15, pp. 3931–3939, 2007.

[18] C. Vannahme, S. Klinkhammer, A. Kolew, P.-J. Jakobs, M. Guttmann, S. Dehm, U. Lemmer, and T. Mappes, "Integration of organic semiconductor lasers and single-mode passive waveguides into a PMMA substrate," *Microelectronic Engineering*, vol. 87, pp. 693–695, 2010.

[19] A. E. Vasdekis, G. Tsiminis, J.-C. Ribierre, L. O' Faolain, T. F. Krauss, G. A. Turnbull, and I. D. W. Samuel, "Diode pumped distributed Bragg reflector lasers based on a dye-to-polymer energy transfer blend," *Optics Express*, vol. 14, pp. 9211–9216, 2006.

[20] T. Riedl, T. Rabe, H.-H. Johannes, W. Kowalsky, J. Wang, T. Weimann, P. Hinze, B. Nehls, T. Farrell, and U. Scherf, "Tunable organic thin-film laser pumped by an inorganic violet diode laser," *Applied Physics Letters*, vol. 88, p. 241116, 2006.

[21] C. Karnutsch, M. Stroisch, M. Punke, U. Lemmer, J. Wang, and T. Weimann, "Laser diode-pumped organic semiconductor lasers utilizing two-dimensional photonic crystal resonators," *Photonics Technology Letters, IEEE*, vol. 19, pp. 741–743, 2007.

[22] S. Klinkhammer, X. Liu, K. Huska, Y. Shen, S. Vanderheiden, S. Valouch, C. Vannahme, S. Bräse, T. Mappes, and U. Lemmer, "Continuously tunable solution-processed organic semiconductor DFB lasers pumped by laser diode," *Optics Express*, vol. 20, pp. 6357–6364, 2012.

[23] Y. Yang, G. A. Turnbull, and I. D. W. Samuel, "Hybrid optoelectronics: A polymer laser pumped by a nitride light-emitting diode," *Applied Physics Letters*, vol. 92, p. 163306, 2008.

[24] Y. Wang, P. O. Morawska, A. L. Kanibolotsky, P. J. Skabara, G. A. Turnbull, and I. D. W. Samuel, "LED pumped polymer laser sensor for explosives," *Laser & Photonics Reviews*, vol. 7, pp. L71–L76, 2013.

[25] J. Herrnsdorf, Y. Wang, J. J. D. McKendry, Z. Gong, D. Massoubre, B. Guilhabert, G. Tsiminis, G. A. Turnbull, I. D. W. Samuel, N. Laurand, E. Gu, and M. D. Dawson, "Micro-LED pumped polymer laser: A discussion of future pump sources for organic lasers," *Laser & Photonics Reviews*, vol. 7, pp. 1065–1078, 2013.

[26] G. Tsiminis, Y. Wang, A. L. Kanibolotsky, A. R. Inigo, P. J. Skabara, I. D. W. Samuel, and G. A. Turnbull, "Nanoimprinted organic semiconductor laser pumped by a light-emitting diode," *Advanced Materials*, vol. 25, pp. 2826–2830, 2013.

[27] D. Schneider, T. Rabe, T. Riedl, T. Dobbertin, M. Kröger, E. Becker, H.-H. Johannes, W. Kowalsky, T. Weimann, J. Wang, P. Hinze, A. Gerhard, P. Stössel, and H. Vestweber, "An ultraviolet organic thin-film solid-state laser for biomarker applications," *Advanced Materials*, vol. 17, pp. 31–34, 2005.

[28] Y. Oki, S. Miyamoto, M. Maeda, and N. J. Vasa, "Multiwavelength distributed-feedback dye laser array and its application to spectroscopy," *Optics Letters*, vol. 27, pp. 1220–1222, 2002.

[29] T. Woggon, S. Klinkhammer, and U. Lemmer, "Compact spectroscopy system based on tunable organic semiconductor lasers," *Applied Physics B*, vol. 99, pp. 47–51, 2010.

[30] C. Vannahme, S. Klinkhammer, U. Lemmer, and T. Mappes, "Plastic lab-on-a-chip for fluorescence excitation with integrated organic semiconductor lasers," *Optics Express*, vol. 19, pp. 8179–8186, 2011.

[31] H. Haken and H. C. Wolf, Eds., *Molekülphysik und Quantenchemie: Einführung in die experimentellen und theoretischen Grundlagen*. Berlin: Springer, 2006.

[32] P. Y. Bruice, *Organic Chemistry*. Upper-Saddle-River: Prentice Hall, 2004.

[33] C. W. Tang and S. A. Vanslyke, "Organic electroluminescent diodes," *Applied Physics Letters*, vol. 51, pp. 913–915, 1987.

[34] M. Cai, T. Xiao, E. Hellerich, Y. Chen, R. Shinar, and J. Shinar, "High-efficiency solution-processed small molecule electrophosphorescent organic light-emitting diodes," *Advanced Materials*, vol. 23, pp. 3590–3596, 2011.

[35] Y. Sun, G. C. Welch, W. L. Leong, C. J. Takacs, G. C. Bazan, and A. J. Heeger, "Solution-processed small-molecule solar cells with 6.7% efficiency," *Nature Materials*, vol. 11, pp. 44–48, 2012.

[36] G. Heliotis, D. D. C. Bradley, G. A. Turnbull, and I. D. W. Samuel, "Light amplification and gain in polyfluorene waveguides," *Applied Physics Letters*, vol. 81, pp. 415–417, 2002.

[37] M. Theander, T. Granlund, D. M. Johanson, A. Ruseckas, V. Sundström, M. R. Andersson, and O. Inganäs, "Lasing in a microcavity with an oriented liquid-crystalline polyfluorene copolymer as active layer," *Advanced Materials*, vol. 13, pp. 323–327, 2001.

[38] U. Scherf and E. List, "Semiconducting polyfluorenes-towards reliable structure-property relationships," *Advanced Materials*, vol. 14, pp. 477–487, 2002.

[39] R. Xia, G. Heliotis, P. N. Stavrinou, and D. D. C. Bradley, "Polyfluorene distributed feedback lasers operating in the green-yellow spectral region," *Applied Physics Letters*, vol. 87, pp. 031 104–3, 2005.

[40] S. V. Frolov, A. Fujii, D. Chinn, M. Hirohata, R. Hidayat, M. Taraguchi, T. Masuda, K. Yoshino, and Z. V. Vardeny, "Microlasers and micro-LEDs from disubstituted polyacetylene," *Advanced Materials*, vol. 10, pp. 869–872, 1998.

[41] N. Johansson, J. Salbeck, J. Bauer, F. Weissörtel, P. Bröms, A. Andersson, and W. Salaneck, "Solid-state amplified spontaneous emission in some spiro-type molecules: A new concept for the design of solid-state lasing molecules," *Advanced Materials*, vol. 10, pp. 1136–1141, 1998.

[42] T. Spehr, A. Siebert, T. Fuhrmann-Lieker, J. Salbeck, T. Rabe, T. Riedl, H. H. Johannes, W. Kowalsky, J. Wang, T. Weimann, and P. Hinze, "Organic solid-state ultraviolet-laser based on spiroterphenyl," *Applied Physics Letters*, vol. 87, p. 161103, 2005.

[43] R. Xia, G. Heliotis, M. Campoy-Quiles, P. N. Stavrinou, D. D. C. Bradley, D. Vak, and D.-Y. Kim, "Characterization of a high-thermal-stability spiroanthracenefluorene-based blue-light-emitting polymer optical gain medium," *Journal of Applied Physics*, vol. 98, p. 083101, 2005.

[44] J. R. Lawrence, G. A. Turnbull, I. D. W. Samuel, G. J. Richards, and P. L. Burn, "Optical amplification in a first-generation dendritic organic semiconductor," *Optics Letters*, vol. 29, pp. 869–871, 2004.

[45] U. Lemmer, *Lecture "Plastic Electronics"*, Light Technology Institute, Karlsruhe Institute of Technology, 2008.

[46] G. M. Barrow, *Introduction to molecular spectroscopy*. New York: MacGraw-Hill, 1962.

[47] M. Pope and C. E. Swenberg, *Electronic processes in organic crystals and polymers*. New York: Oxford University Press, 1999.

[48] P. W. Atkins and R. Friedman, *Molecular quantum mechanics*. Oxford: Oxford University Press, 2011.

[49] C. Karnutsch, "Low threshold organic thin-film laser devices," Ph.D. dissertation, Light Technology Institute, Karlsruhe Institute of Technology, 2007.

[50] R. Enderlein and N. J. M. Horing, *Fundamentals of semiconductor physics and devices*. Singapore: World Scientific, 1997.

[51] P. Y. Yu and M. Cardona, *Fundamentals of semiconductors : physics and materials properties*. Berlin: Springer, 2005.

[52] V. Coropceanu, J. Cornil, D. A. da Silva Filho, Y. Olivier, R. Silbey, and J.-L. Brédas, "Charge transport in organic semiconductors," *Chemical Reviews*, vol. 107, pp. 926–952, 2007.

[53] S. Riechel, "Organic semiconductor lasers with two-dimensional distributed feedback," Ph.D. dissertation, Ludwig Maximilian University of Munich, 2002.

[54] H. Bässler, G. Schönherr, M. Abkowitz, and D. M. Pai, "Hopping transport in prototypical organic glasses," *Physical Review B*, vol. 26, pp. 3105–3113, 1982.

[55] M. Raikh and X. Wei, "Current-voltage characteristics of polymer light-emitting diode at low voltages," *Molecular Crystals and Liquid Crystals Science and Technology. Section A. Molecular Crystals and Liquid Crystals*, vol. 256, pp. 563–569, 1994.

[56] U. Lemmer and E. O. Göbel, "Photoluminescence spectroscopy as a probe for disorder and excitonic effects in organic and inorganic semiconductors," in *Primary Photoexcitations in Conjugated Polymers: Molecular Exciton Versus Semiconductor Band Model*, N. S. Sariciftci, Ed. Singapore: World Scientific, 1997, ch. 9, pp. 211–253.

[57] D. R. Kearns, G. Tollin, and M. Calvin, "Electrical properties of organic solids. II. Effects of added electron acceptor on metal-free

phthalocyanine," *The Journal of Chemical Physics*, vol. 32, pp. 1020–1025, 1960.

[58] M. A. Lampert and P. Mark, *Current injection in solids*. New York: Academic Press, 1970.

[59] C. K. Chiang, C. R. Fincher, Y. W. Park, A. J. Heeger, H. Shirakawa, E. J. Louis, S. C. Gau, and A. G. MacDiarmid, "Electrical conductivity in doped polyacetylene," *Physical Review Letters*, vol. 39, pp. 1098–1101, 1977.

[60] M. Koehler and I. A. Hümmelgen, "Regional approximation approach to space charge limited tunneling injection in polymeric devices," *Journal of Applied Physics*, vol. 87, pp. 3074–3079, 2000.

[61] J. Endo, T. Matsumoto, and J. Kido, "Organic electroluminescent devices with a vacuum-deposited Lewis-acid-doped hole-injecting layer," *Japanese Journal of Applied Physics*, vol. 41, pp. L358–L360, 2002.

[62] Z. J. Wang, Y. Wu, Y. C. Zhou, J. Zhou, S. T. Zhang, X. M. Ding, X. Y. Hou, and Z. Q. Zhu, "Control of carrier transport in organic semiconductors by aluminum doping," *Applied Physics Letters*, vol. 88, p. 222112, 2006.

[63] A. Werner, F. Li, K. Harada, M. Pfeiffer, T. Fritz, K. Leo, and S. Machill, "n-type doping of organic thin films using cationic dyes," *Advanced Functional Materials*, vol. 14, pp. 255–260, 2004.

[64] A. E. Siegman, *Lasers*. Sausalito: Univ Science Books, 1986.

[65] B. E. A. Saleh and M. C. Teich, *Fundamentals of photonics*. Hoboken: Wiley, 2007.

[66] I. D. W. Samuel, E. B. Namdas, and G. A. Turnbull, "How to recognize lasing," *Nature Photonics*, vol. 3, pp. 546–549, 2009.

[67] F. P. H. Schäfer and K. H. M. Drexhage, Eds., *Dye lasers*. Berlin: Springer, 1990.

[68] I. D. W. Samuel and G. A. Turnbull, "Organic semiconductor lasers," *Chemical Reviews*, vol. 107, pp. 1272–1295, 2007.

[69] S. Forget and S. Chénais, *Organic solid-state lasers*. Berlin: Springer, 2013.

[70] F. J. H. Duarte, Ed., *High-power dye lasers*. Berlin: Springer, 1991.

[71] D. Amarasinghe, A. Ruseckas, A. E. Vasdekis, M. Goossens, G. A. Turnbull, and I. D. W. Samuel, "Broadband solid state optical amplifier based on a semiconducting polymer," *Applied Physics Letters*, vol. 89, p. 201119, 2006.

[72] K. L. Shaklee and R. F. Leheny, "Direct determination of optical gain in semiconductor crystals," *Applied Physics Letters*, vol. 18, pp. 475–477, 1971.

[73] M. D. McGehee, R. Gupta, S. Veenstra, E. K. Miller, M. A. Díaz-García, and A. J. Heeger, "Amplified spontaneous emission from photopumped films of a conjugated polymer," *Physical Review B*, vol. 58, pp. 7035–7039, 1998.

[74] R. Xia, G. Heliotis, Y. Hou, and D. D. Bradley, "Fluorene-based conjugated polymer optical gain media," *Organic Electronics*, vol. 4, pp. 165–177, 2003.

[75] G. Jordan, M. Flämmich, M. Rüther, T. Kobayashi, W. J. Blau, Y. Suzuki, and T. Kaino, "Light amplification at 501nm and large nanosecond optical gain in organic dye-doped polymeric waveguides," *Applied Physics Letters*, vol. 88, p. 161114, 2006.

[76] I. Samuel, F. Raksi, D. Bradley, R. Friend, P. Burn, A. Holmes, H. Murata, T. Tsutsui, and S. Saito, "Femtosecond transient absorption measurements in poly(arylenevinylene)s," *Synthetic Metals*, vol. 55, pp. 15–21, 1993.

[77] M. Yan, L. J. Rothberg, F. Papadimitrakopoulos, M. E. Galvin, and T. M. Miller, "Spatially indirect excitons as primary photoexcitations in conjugated polymers," *Physical Review Letters*, vol. 72, pp. 1104–1107, 1994.

[78] G. J. Denton, N. Tessler, N. T. Harrison, and R. H. Friend, "Factors influencing stimulated emission from poly(p-phenylenevinylene)," *Physical Review Letters*, vol. 78, pp. 733–736, 1997.

[79] S. V. Frolov, Z. V. Vardeny, and K. Yoshino, "Cooperative and stimulated emission in poly(p-phenylene-vinylene) thin films and solutions," *Physical Review B*, vol. 57, pp. 9141–9147, 1998.

[80] G. Cerullo, S. Stagira, M. Nisoli, S. De Silvestri, G. Lanzani, G. Kranzelbinder, W. Graupner, and G. Leising, "Excited-state dynamics of poly(para-phenylene)-type ladder polymers at high photoexcitation density," *Physical Review B*, vol. 57, pp. 12 806–12 811, 1998.

[81] B. Kraabel, V. I. Klimov, R. Kohlman, S. Xu, H.-L. Wang, and D. W. McBranch, "Unified picture of the photoexcitations in phenylene-based conjugated polymers: Universal spectral and dynamical features in subpicosecond transient absorption," *Physical Review B*, vol. 61, pp. 8501–8515, 2000.

[82] M. Punke, F. Hoos, C. Karnutsch, U. Lemmer, N. Linder, and K. Streubel, "High-repetition-rate white-light pump-probe spectroscopy with a tapered fiber," *Optics Letters*, vol. 31, pp. 1157–1159, 2006.

[83] M. A. Díaz-García, F. Hide, B. J. Schwartz, M. R. Andersson, Q. Pei, and A. J. Heeger, "Plastic lasers: Semiconducting polymers as a new class of solid-state laser materials," *Synthetic Metals*, vol. 84, pp. 455–462, 1997.

[84] G. Kranzelbinder and G. Leising, "Organic solid-state lasers," *Reports on Progress in Physics*, vol. 63, pp. 729–762, 2000.

[85] S. Riechel, U. Lemmer, J. Feldmann, S. Berleb, A. G. Mückl, W. Brütting, A. Gombert, and V. Wittwer, "Very compact tunable solid-state laser utilizing a thin-film organic semiconductor," *Optics Letters*, vol. 26, pp. 593–595, 2001.

[86] J. E. Geusic, H. M. Marcos, and L. G. Van Uitert, "Laser oscillations in Nd-doped yttrium aluminum, yttrium gallium and gadolinium garnets," *Applied Physics Letters*, vol. 4, pp. 182–184, 1964.

[87] Photonics Handbook, "Nd:YAG lasers," http://photonics.com/edu/Handbook.aspx, 2014.

[88] G. Morthier and P. Vankwikelberge, *Handbook of distributed feedback laser diodes*. Boston: Artech House, 1997.

[89] M. Reufer, J. M. Lupton, and U. Scherf, "Stimulated emission depletion of triplet excitons in a phosphorescent organic laser," *Applied Physics Letters*, vol. 89, p. 141111, 2006.

[90] M. Lehnhardt, T. Riedl, T. Rabe, and W. Kowalsky, "Room temperature lifetime of triplet excitons in fluorescent host/guest systems," *Organic Electronics*, vol. 12, pp. 486–491, 2011.

[91] M. Yamashita and H. Kashiwagi, "Photodegradation mechanisms in laser dyes: A laser irradiated ESR study," *IEEE Journal of Quantum Electronics*, vol. 12, pp. 90–95, 1976.

[92] B. H. Cumpston and K. F. Jensen, "Photo-oxidation of polymers used in electroluminescent devices," *Synthetic Metals*, vol. 73, pp. 195–199, 1995.

[93] R. D. Scurlock, B. Wang, P. R. Ogilby, J. R. Sheats, and R. L. Clough, "Singlet oxygen as a reactive intermediate in the photodegradation of an electroluminescent polymer," *Journal of the American Chemical Society*, vol. 117, pp. 10 194–10 202, 1995.

[94] J. Yu, D. Hu, and P. F. Barbara, "Unmasking electronic energy transfer of conjugated polymers by suppression of O_2 quenching," *Science*, vol. 289, pp. 1327–1330, 2000.

[95] C.-L. Lee, X. Yang, and N. C. Greenham, "Determination of the triplet excited-state absorption cross section in a polyfluorene by energy transfer from a phosphorescent metal complex," *Physical Review B*, vol. 76, p. 245201, 2007.

[96] C. Gaertner, C. Karnutsch, U. Lemmer, and C. Pflumm, "The influence of annihilation processes on the threshold current density of organic laser diodes," *Journal of Applied Physics*, vol. 101, p. 023107, 2007.

[97] M. Lehnhardt, T. Riedl, T. Weimann, and W. Kowalsky, "Impact of triplet absorption and triplet-singlet annihilation on the dynamics

of optically pumped organic solid-state lasers," *Physical Review B*, vol. 81, p. 165206, 2010.

[98] S. B. Klinkhammer, "Durchstimmbare organische Halbleiterlaser," Ph.D. dissertation, Karlsruhe Institute of Technology, Karlsruhe, 2012.

[99] X. Yang, C.-L. Lee, S. Westenhoff, X. Zhang, and N. C. Greenham, "Saturation, relaxation, and dissociation of excited triplet excitons in conjugated polymers," *Advanced Materials*, vol. 21, pp. 916–919, 2009.

[100] S.-L. Chua, B. Zhen, J. Lee, J. Bravo-Abad, O. Shapira, and M. Soljacic, "Modeling of threshold and dynamics behavior of organic nanostructured lasers," *Journal of Materials Chemistry C*, vol. 2, pp. 1463–1473, 2014.

[101] T. Rabe, K. Gerlach, T. Riedl, H.-H. Johannes, W. Kowalsky, J. Niederhofer, W. Gries, J. Wang, T. Weimann, P. Hinze, F. Galbrecht, and U. Scherf, "Quasi-continuous-wave operation of an organic thin-film distributed feedback laser," *Applied Physics Letters*, vol. 89, p. 081115, 2006.

[102] M. Lehnhardt, T. Riedl, U. Scherf, T. Rabe, and W. Kowalsky, "Spectrally separated optical gain and triplet absorption: Towards continuous wave lasing in organic thin film lasers," *Organic Electronics*, vol. 12, pp. 1346–1351, 2011.

[103] R. Bornemann, U. Lemmer, and E. Thiel, "Continuous-wave solid-state dye laser," *Optics Letters*, vol. 31, pp. 1669–1671, 2006.

[104] S. Schols, A. Kadashchuk, P. Heremans, A. Helfer, and U. Scherf, "Triplet excitation scavenging in films of conjugated polymers," *ChemPhysChem*, vol. 10, pp. 1071–1076, 2009.

[105] Y. Zhang and S. R. Forrest, "Existence of continuous-wave threshold for organic semiconductor lasers," *Physical Review B*, vol. 84, p. 241301, 2011.

[106] T. Förster, "Zwischenmolekulare Energiewanderung und Fluoreszenz," *Annals of Physics*, vol. 437, pp. 55–75, 1948.

[107] M. Berggren, A. Dodabalapur, and R. E. Slusher, "Stimulated emission and lasing in dye-doped organic thin films with Forster transfer," *Applied Physics Letters*, vol. 71, pp. 2230–2232, 1997.

[108] R. Gupta, M. Stevenson, and A. J. Heeger, "Low threshold distributed feedback lasers fabricated from blends of conjugated polymers: Reduced losses through Förster transfer," *Journal of Applied Physics*, vol. 92, pp. 4874–4877, 2002.

[109] V. Bulovic, V. G. Kozlov, V. B. Khalfin, and S. R. Forrest, "Transform-limited, narrow-linewidth lasing action in organic semiconductor microcavities," *Science*, vol. 279, pp. 553–555, 1998.

[110] R. G. Kepler, P. M. Beeson, S. J. Jacobs, R. A. Anderson, M. B. Sinclair, V. S. Valencia, and P. A. Cahill, "Electron and hole mobility in tris(8-hydroxyquinolinolato-n1,o8) aluminum," *Applied Physics Letters*, vol. 66, pp. 3618–3620, 1995.

[111] H. Aziz, Z. Popovic, S. Xie, A.-M. Hor, N.-X. Hu, C. Tripp, and G. Xu, "Humidity-induced crystallization of tris (8-hydroxyquinoline) aluminum layers in organic light-emitting devices," *Applied Physics Letters*, vol. 72, pp. 756–758, 1998.

[112] V. G. Kozlov, V. Bulovic, P. E. Burrows, M. Baldo, V. B. Khalfin, G. Parthasarathy, S. R. Forrest, Y. You, and M. E. Thompson, "Study of lasing action based on Förster energy transfer in optically pumped organic semiconductor thin films," *Journal of Applied Physics*, vol. 84, pp. 4096–4108, 1998.

[113] D. Schneider, T. Rabe, T. Riedl, T. Dobbertin, M. Kröger, E. Becker, H.-H. Johannes, W. Kowalsky, T. Weimann, J. Wang, and P. Hinze, "Ultrawide tuning range in doped organic solid-state lasers," *Applied Physics Letters*, vol. 85, pp. 1886–1888, 2004.

[114] V. Bulovic, A. Shoustikov, M. Baldo, E. Bose, V. Kozlov, M. Thompson, and S. Forrest, "Bright, saturated, red-to-yellow organic light-emitting devices based on polarization-induced spectral shifts," *Chemical Physics Letters*, vol. 287, pp. 455–460, 1998.

[115] V. Bulovic, R. Deshpande, M. Thompson, and S. Forrest, "Tuning the color emission of thin film molecular organic light emitting devices by the solid state solvation effect," *Chemical Physics Letters*, vol. 308, pp. 317–322, 1999.

[116] S. Riechel, U. Lemmer, J. Feldmann, T. Benstem, W. Kowalsky, U. Scherf, A. Gombert, and V. Wittwer, "Laser modes in organic solid-state distributed feedback lasers," *Applied Physics B*, vol. 71, pp. 897–900, 2000.

[117] N. Tsutsumi, A. Fujihara, and D. Hayashi, "Tunable distributed feedback lasing with a threshold in the nanojoule range in an organic guest-host polymeric waveguide," *Applied Optics*, vol. 45, pp. 5748–5751, 2006.

[118] V. Kozlov, G. Parthasarathy, P. E. Burrows, V. Khalfin, J. Wang, S. Chou, and S. Forrest, "Structures for organic diode lasers and optical properties of organic semiconductors under intense optical and electrical excitations," *IEEE Journal of Quantum Electronics*, vol. 36, pp. 18–26, 2000.

[119] J. Wang, T. Weimann, P. Hinze, G. Ade, D. Schneider, T. Rabe, T. Riedl, and W. Kowalsky, "A continuously tunable organic DFB laser," *Microelectronic Engineering*, vol. 78–79, pp. 364–368, 2005.

[120] M. Punke, T. Woggon, M. Stroisch, B. Ebenhoch, U. Geyer, C. Karnutsch, M. Gerken, U. Lemmer, M. Bruendel, J. Wang, and T. Weimann, "Organic semiconductor lasers as integrated light sources for optical sensor systems," in *Proc. SPIE*, vol. 6659, p. 665909, 2007.

[121] S. Klinkhammer, T. Woggon, U. Geyer, C. Vannahme, S. Dehm, T. Mappes, and U. Lemmer, "A continuously tunable low-threshold organic semiconductor distributed feedback laser fabricated by rotating shadow mask evaporation," *Applied Physics B*, vol. 97, pp. 787–791, 2009.

[122] American Dye Source, Inc., "Products: OLED&PLED," http://www.adsdyes.com/oled.html, 2010.

[123] M. A. Stevens, C. Silva, D. M. Russell, and R. H. Friend, "Exciton dissociation mechanisms in the polymeric semiconductors poly(9,9-dioctylfluorene) and poly(9,9-dioctylfluorene-co-benzothiadiazole)," *Physical Review B*, vol. 63, p. 165213, 2001.

[124] C. M. Ramsdale and N. C. Greenham, "The optical constants of emitter and electrode materials in polymer light-emitting diodes," *Journal of Physics D: Applied Physics*, vol. 36, pp. L29–L34, 2003.

[125] M. Vehse, B. Liu, L. Edman, G. Bazan, and A. Heeger, "Light amplification by optical excitation of a chemical defect in a conjugated polymer," *Advanced Materials*, vol. 16, pp. 1001–1004, 2004.

[126] D. Moses, "High quantum efficiency luminescence from a conducting polymer in solution: A novel polymer laser dye," *Applied Physics Letters*, vol. 60, pp. 3215–3216, 1992.

[127] L. J. Rothberg, M. Yan, F. Papadimitrakopoulos, M. E. Galvin, E. W. Kwock, and T. M. Miller, "Photophysics of phenylenevinylene polymers," *Synthetic Metals*, vol. 80, pp. 41–58, 1996.

[128] T.-Q. Nguyen, V. Doan, and B. J. Schwartz, "Conjugated polymer aggregates in solution: Control of interchain interactions," *The Journal of Chemical Physics*, vol. 110, pp. 4068–4078, 1999.

[129] M. D. McGehee and A. J. Heeger, "Semiconducting (conjugated) polymers as materials for solid-state lasers," *Advanced Materials*, vol. 12, pp. 1655–1668, 2000.

[130] F. Hide, B. J. Schwartz, M. A. Díaz-García, and A. J. Heeger, "Laser emission from solutions and films containing semiconducting polymer and titanium dioxide nanocrystals," *Chemical Physics Letters*, vol. 256, pp. 424–430, 1996.

[131] S. Chénais and S. Forget, "Recent advances in solid-state organic lasers," *Polymer International*, vol. 61, pp. 390–406, 2012.

[132] L. El-Nadi, L. Al-Houty, M. Omar, and M. Ragab, "Organic thin film materials producing novel blue laser," *Chemical Physics Letters*, vol. 286, pp. 9–14, 1998.

[133] J. H. Schön, C. Kloc, A. Dodabalapur, and B. Batlogg, "An organic solid state injection laser," *Science*, vol. 289, pp. 599–601, 2000.

[134] D. Yokoyama, M. Moriwake, and C. Adachi, "Spectrally narrow emissions at cutoff wavelength from edges of optically and electrically pumped anisotropic organic films," *Journal of Applied Physics*, vol. 103, p. 123104, 2008.

[135] Y. Tian, Z. Gan, Z. Zhou, D. W. Lynch, J. Shinar, J.-h. Kang, and Q.-H. Park, "Spectrally narrowed edge emission from organic light-emitting diodes," *Applied Physics Letters*, vol. 91, p. 143504, 2007.

[136] X. Liu, H. Li, C. Song, Y. Liao, and M. Tian, "Microcavity organic laser device under electrical pumping," *Optics Letters*, vol. 34, pp. 503–505, 2009.

[137] A. Köhler and H. Bässler, "Triplet states in organic semiconductors," *Materials Science and Engineering: R: Reports*, vol. 66, pp. 71–109, 2009.

[138] P. Andrew, G. A. Turnbull, I. D. W. Samuel, and W. L. Barnes, "Photonic band structure and emission characteristics of a metal-backed polymeric distributed feedback laser," *Applied Physics Letters*, vol. 81, pp. 954–956, 2002.

[139] M. Reufer, S. Riechel, J. M. Lupton, J. Feldmann, U. Lemmer, D. Schneider, T. Benstem, T. Dobbertin, W. Kowalsky, A. Gombert, K. Forberich, V. Wittwer, and U. Scherf, "Low-threshold polymeric distributed feedback lasers with metallic contacts," *Applied Physics Letters*, vol. 84, pp. 3262–3264, 2004.

[140] P. Görrn, T. Rabe, T. Riedl, and W. Kowalsky, "Loss reduction in fully contacted organic laser waveguides using TE_2 modes," *Applied Physics Letters*, vol. 91, p. 041113, 2007.

[141] B. H. Wallikewitz, M. de la Rosa, J. H.-W. M. Kremer, D. Hertel, and K. Meerholz, "A lasing organic light-emitting diode," *Advanced Materials*, vol. 22, pp. 531–534, 2010.

[142] H. Yamamoto, T. Oyamada, H. Sasabe, and C. Adachi, "Amplified spontaneous emission under optical pumping from an organic semiconductor laser structure equipped with transparent carrier injection electrodes," *Applied Physics Letters*, vol. 84, pp. 1401–1403, 2004.

[143] M. A. Díaz-García, F. Hide, B. J. Schwartz, M. D. McGehee, M. R. Andersson, and A. J. Heeger, "'Plastic' lasers: Comparison of gain narrowing with a soluble semiconducting polymer in waveguides and microcavities," *Applied Physics Letters*, vol. 70, pp. 3191–3193, 1997.

[144] T. Virgili, D. G. Lidzey, M. Grell, D. D. C. Bradley, S. Stagira, M. Zavelani-Rossi, and S. De Silvestri, "Influence of the orientation of liquid crystalline poly(9,9-dioctylfluorene) on its lasing properties in a planar microcavity," *Applied Physics Letters*, vol. 80, pp. 4088–4090, 2002.

[145] M. Koschorreck, R. Gehlhaar, V. G. Lyssenko, M. Swoboda, M. Hoffmann, and K. Leo, "Dynamics of a high-Q vertical-cavity organic laser," *Applied Physics Letters*, vol. 87, p. 181108, 2005.

[146] L. Persano, A. Camposeo, P. Del Carro, E. Mele, R. Cingolani, and D. Pisignano, "Low-threshold blue-emitting monolithic polymer vertical cavity surface-emitting lasers," *Applied Physics Letters*, vol. 89, p. 121111, 2006.

[147] A. Schülzgen, C. Spiegelberg, M. M. Morrell, S. B. Mendes, B. Kippelen, N. Peyghambarian, M. F. Nabor, E. A. Mash, and P. M. Allemand, "Near diffraction-limited laser emission from a polymer in a high finesse planar cavity," *Applied Physics Letters*, vol. 72, pp. 269–271, 1998.

[148] S. V. Frolov, M. Shkunov, Z. V. Vardeny, and K. Yoshino, "Ring microlasers from conducting polymers," *Physical Review B*, vol. 56, pp. R4363–R4366, 1997.

[149] S. V. Frolov, Z. V. Vardeny, and K. Yoshino, "Plastic microring lasers on fibers and wires," *Applied Physics Letters*, vol. 72, pp. 1802–1804, 1998.

[150] Y. Kawabe, C. Spiegelberg, A. Schülzgen, M. F. Nabor, B. Kippelen, E. A. Mash, P. M. Allemand, M. Kuwata-Gonokami, K. Takeda, and N. Peyghambarian, "Whispering-gallery-mode microring laser using a conjugated polymer," *Applied Physics Letters*, vol. 72, pp. 141–143, 1998.

[151] R. Österbacka, M. Wohlgenannt, M. Shkunov, D. Chinn, and Z. V. Vardeny, "Excitons, polarons, and laser action in poly(p-phenylene vinylene) films," *The Journal of Chemical Physics*, vol. 118, pp. 8905–8916, 2003.

[152] M. Berggren, A. Dodabalapur, R. Slusher, and Z. Bao, "Organic lasers based on Förster transfer," *Synthetic Metals*, vol. 91, pp. 65–68, 1997.

[153] C. Sheng, R. Polson, Z. Vardeny, and D. Chinn, "Studies of π-conjugated polymer coupled microlasers," *Synthetic Metals*, vol. 135–136, pp. 147–149, 2003.

[154] M. Berggren, A. Dodabalapur, Z. Bao, and R. E. Slusher, "Solid-state droplet laser made from an organic blend with a conjugated polymer emitter," *Advanced Materials*, vol. 9, pp. 968–971, 1997.

[155] R. C. Polson, G. Levina, and Z. V. Vardeny, "Spectral analysis of polymer microring lasers," *Applied Physics Letters*, vol. 76, pp. 3858–3860, 2000.

[156] R. Baktur, L. Pearson, and J. Ballato, "Theoretical determination of lasing resonances in a microring," *Journal of Applied Physics*, vol. 101, p. 043102, 2007.

[157] M. Lebental, N. Djellali, C. Arnaud, J.-S. Lauret, J. Zyss, R. Dubertrand, C. Schmit, and E. Bogomolny, "Inferring periodic orbits from spectra of simply shaped microlasers," *Physical Review A*, vol. 76, p. 023830, 2007.

[158] M. Lebental, J. S. Lauret, J. Zyss, C. Schmit, and E. Bogomolny, "Directional emission of stadium-shaped microlasers," *Physical Review A*, vol. 75, p. 033806, 2007.

[159] J. A. Rogers, M. Meier, and A. Dodabalapur, “Using printing and molding techniques to produce distributed feedback and Bragg reflector resonators for plastic lasers,” *Applied Physics Letters*, vol. 73, pp. 1766–1768, 1998.

[160] M. D. McGehee, M. A. Díaz-García, F. Hide, R. Gupta, E. K. Miller, D. Moses, and A. J. Heeger, “Semiconducting polymer distributed feedback lasers,” *Applied Physics Letters*, vol. 72, pp. 1536–1538, 1998.

[161] C. Kallinger, M. Hilmer, A. Haugeneder, M. Perner, W. Spirkl, U. Lemmer, J. Feldmann, U. Scherf, K. Müllen, A. Gombert, and V. Wittwer, “A flexible conjugated polymer laser,” *Advanced Materials*, vol. 10, pp. 920–923, 1998.

[162] M. Berggren, A. Dodabalapur, R. E. Slusher, A. Timko, and O. Nalamasu, “Organic solid-state lasers with imprinted gratings on plastic substrates,” *Applied Physics Letters*, vol. 72, pp. 410–411, 1998.

[163] G. Heliotis, R. Xia, D. D. C. Bradley, G. A. Turnbull, I. D. W. Samuel, P. Andrew, and W. L. Barnes, “Blue, surface-emitting, distributed feedback polyfluorene lasers,” *Applied Physics Letters*, vol. 83, pp. 2118–2120, 2003.

[164] A. Dodabalapur, M. Berggren, R. Slusher, Z. Bao, A. Timko, P. Schiortino, E. Laskowski, H. Katz, and O. Nalamasu, “Resonators and materials for organic lasers based on energy transfer,” *Journal of Selected Topics in Quantum Electronics*, vol. 4, pp. 67–74, 1998.

[165] M. Notomi, H. Suzuki, T. Tamamura, and K. Edagawa, “Lasing action due to the two-dimensional quasiperiodicity of photonic quasicrystals with a penrose lattice,” *Physical Review Letters*, vol. 92, p. 123906, 2004.

[166] G. Heliotis, R. Xia, G. A. Turnbull, P. Andrew, W. L. Barnes, I. W. Samuel, and D. D. C. Bradley, “Emission characteristics and performance comparison of polyfluorene lasers with one- and two-dimensional distributed feedback,” *Advanced Functional Materials*, vol. 14, pp. 91–97, 2004.

[167] A. Yariv, *Optical electronics in modern communications*. Oxford: Oxford University Press, 1997.

[168] A. Yariv and M. Nakamura, "Periodic structures for integrated optics," *IEEE Journal of Quantum Electronics*, vol. 13, pp. 233–253, 1977.

[169] S. Riechel, C. Kallinger, U. Lemmer, J. Feldmann, A. Gombert, V. Wittwer, and U. Scherf, "A nearly diffraction limited surface emitting conjugated polymer laser utilizing a two-dimensional photonic band structure," *Applied Physics Letters*, vol. 77, pp. 2310–2312, 2000.

[170] A. E. Vasdekis, G. A. Turnbull, I. D. W. Samuel, P. Andrew, and W. L. Barnes, "Low threshold edge emitting polymer distributed feedback laser based on a square lattice," *Applied Physics Letters*, vol. 86, p. 161102, 2005.

[171] C. Karnutsch, C. Gärtner, V. Haug, U. Lemmer, T. Farrell, B. S. Nehls, U. Scherf, J. Wang, T. Weimann, G. Heliotis, C. Pflumm, J. C. deMello, and D. D. C. Bradley, "Low threshold blue conjugated polymer lasers with first- and second-order distributed feedback," *Applied Physics Letters*, vol. 89, p. 201108, 2006.

[172] J. E. Carroll, J. Whiteaway, and D. Plumb, *Distributed feedback semiconductor lasers*. London: Institution of Electrical Engineers, 1998.

[173] C. Karnutsch, C. Pflumm, G. Heliotis, J. C. deMello, D. D. C. Bradley, J. Wang, T. Weimann, V. Haug, C. Gärtner, and U. Lemmer, "Improved organic semiconductor lasers based on a mixed-order distributed feedback resonator design," *Applied Physics Letters*, vol. 90, p. 131104, 2007.

[174] K. Baumann, T. Stäferle, N. Moll, R. F. Mahrt, T. Wahlbrink, J. Bolten, T. Mollenhauer, C. Moormann, and U. Scherf, "Organic mixed-order photonic crystal lasers with ultrasmall footprint," *Applied Physics Letters*, vol. 91, p. 171108, 2007.

[175] H. Kogelnik and C. V. Shank, "Coupled-wave theory of distributed feedback lasers," *Journal of Applied Physics*, vol. 43, pp. 2327–2335, 1972.

[176] W. Streifer, D. Scifres, and R. Burnham, "Coupled wave analysis of DFB and DBR lasers," *IEEE Journal of Quantum Electronics*, vol. 13, pp. 134–141, 1977.

[177] H. Haus and C. Shank, "Antisymmetric taper of distributed feedback lasers," *IEEE Journal of Quantum Electronics*, vol. 12, pp. 532–539, 1976.

[178] E. Kapon, A. Hardy, and A. Katzir, "The effect of complex coupling coefficients on distributed feedback lasers," *IEEE Journal of Quantum Electronics*, vol. 18, pp. 66–71, 1982.

[179] X. Liu, P. Stefanou, B. Wang, T. Woggon, T. Mappes, and U. Lemmer, "Organic semiconductor distributed feedback (DFB) laser as excitation source in Raman spectroscopy," *Optics Express*, vol. 21, pp. 28 941–28 947, 2013.

[180] K. Forberich, M. Diem, J. Crewett, U. Lemmer, A. Gombert, and K. Busch, "Lasing action in two-dimensional organic photonic crystal lasers with hexagonal symmetry," *Applied Physics B: Lasers and Optics*, vol. 82, pp. 539–541, 2006.

[181] C. Bauer, H. Giessen, B. Schnabel, E.-B. Kley, C. Schmitt, U. Scherf, and R. F. Mahrt, "A surface-emitting circular grating polymer laser," *Advanced Materials*, vol. 13, pp. 1161–1164, 2001.

[182] G. A. Turnbull, A. Carleton, G. F. Barlow, A. Tahraouhi, T. F. Krauss, K. A. Shore, and I. D. W. Samuel, "Influence of grating characteristics on the operation of circular-grating distributed-feedback polymer lasers," *Journal of Applied Physics*, vol. 98, p. 023105, 2005.

[183] V. A. Sychugov, A. V. Tishchenko, and A. A. Khakimov, "Two-dimensional periodic structures in thin-film lasers," *Soviet Journal of Quantum Electronics*, vol. 12, pp. 27–29, 1982.

[184] M. Toda, "Proposed cross grating single-mode DFB laser," *IEEE Journal of Quantum Electronics*, vol. 28, pp. 1653–1662, 1992.

[185] H. Han and J. Coleman, "Two-dimensional rectangular lattice distributed feedback lasers: A coupled-mode analysis of TE guided modes," *IEEE Journal of Quantum Electronics*, vol. 31, pp. 1947–1954, 1995.

[186] G. A. Turnbull, P. Andrew, W. L. Barnes, and I. D. W. Samuel, "Photonic mode dispersion of a two-dimensional distributed feedback polymer laser," *Physical Review B*, vol. 67, p. 165107, 2003.

[187] K. Sakai, E. Miyai, and S. Noda, "Coupled-wave theory for square-lattice photonic crystal lasers with TE polarization," *IEEE Journal of Quantum Electronics*, vol. 46, pp. 788–795, 2010.

[188] G. Heliotis, R. Xia, D. Bradley, G. Turnbull, I. Samuel, P. Andrew, and W. Barnes, "Two-dimensional distributed feedback lasers using a broadband, red polyfluorene gain medium," *Journal of Applied Physics*, vol. 96, pp. 6959–6965, 2004.

[189] M. Lu, S. S. Choi, U. Irfan, and B. T. Cunningham, "Plastic distributed feedback laser biosensor," *Applied Physics Letters*, vol. 93, p. 111113, 2008.

[190] M. Lu, S. S. Choi, C. J. Wagner, J. G. Eden, and B. T. Cunningham, "Label free biosensor incorporating a replica-molded, vertically emitting distributed feedback laser," *Applied Physics Letters*, vol. 92, p. 261502, 2008.

[191] A. Rose, Z. Zhu, C. F. Madigan, T. M. Swager, and V. Bulovic, "Sensitivity gains in chemosensing by lasing action in organic polymers," *Nature*, vol. 434, pp. 876–879, 2005.

[192] Y. Yang, G. A. Turnbull, and I. D. W. Samuel, "Sensitive explosive vapor detection with polyfluorene lasers," *Advanced Functional Materials*, vol. 20, pp. 2093–2097, 2010.

[193] Y. Wang, N. B. McKeown, K. J. Msayib, G. A. Turnbull, and I. D. W. Samuel, "Laser chemosensor with rapid responsivity and inherent memory based on a polymer of intrinsic microporosity," *Sensors*, vol. 11, pp. 2478–2487, 2011.

[194] G. Heliotis, D. D. C. Bradley, M. Goossens, S. Richardson, G. A. Turnbull, and I. D. W. Samuel, "Operating characteristics of a traveling-wave semiconducting polymer optical amplifier," *Applied Physics Letters*, vol. 85, pp. 6122–6124, 2004.

[195] D. Amarasinghe, A. Ruseckas, G. Turnbull, and I. D. W. Samuel, "Organic semiconductor optical amplifiers," *Proceedings of the IEEE*, vol. 97, pp. 1637–1650, 2009.

[196] S. V. Frolov, M. Liess, P. A. Lane, W. Gellermann, Z. V. Vardeny, M. Ozaki, and K. Yoshino, "Exciton dynamics in soluble poly(p-phenylene-vinylene): Towards an ultrafast excitonic switch," *Physical Review Letters*, vol. 78, pp. 4285–4288, 1997.

[197] T. Virgili, D. Marinotto, G. Lanzani, and D. D. Bradley, "Ultrafast resonant optical switching in isolated polyfluorenes chains," *Applied Physics Letters*, vol. 86, p. 091113, 2005.

[198] T. Voss, D. Scheel, and W. Schade, "A microchip-laser-pumped DFB-polymer-dye laser," *Applied Physics B*, vol. 73, pp. 105–109, 2001.

[199] X. Zhu and D. Lo, "Temperature tuning of output wavelength for solid-state dye lasers," *Journal of Optics A: Pure and Applied Optics*, vol. 3, p. 225, 2001.

[200] T. Woggon, "Organic solid state lasers for sensing applications," Ph.D. dissertation, Karlsruhe Institute of Technology, 2011.

[201] G. Turnbull, T. Krauss, W. Barnes, and I. Samuel, "Tuneable distributed feedback lasing in MEH-PPV films," *Synthetic Metals*, vol. 121, pp. 1757–1758, 2001.

[202] K. Suzuki, K. Takahashi, Y. Seida, K. Shimizu, M. Kumagai, and Y. Taniguchi, "A continuously tunable organic solid-state laser based on a flexible distributed-feedback resonator," *Japanese Journal of Applied Physics*, vol. 42, pp. L249–L251, 2003.

[203] M. R. Weinberger, G. Langer, A. Pogantsch, A. Haase, E. Zojer, and W. Kern, "Continuously color-tunable rubber laser," *Advanced Materials*, vol. 16, pp. 130–133, 2004.

[204] B. Wenger, N. Tetreault, M. E. Welland, and R. H. Friend, "Mechanically tunable conjugated polymer distributed feedback lasers," *Applied Physics Letters*, vol. 97, p. 193303, 2010.

[205] P. Görrn, M. Lehnhardt, W. Kowalsky, T. Riedl, and S. Wagner, "Elastically tunable self-organized organic lasers," *Advanced Materials*, vol. 23, pp. 869–872, 2011.

[206] S. Döring, M. Kollosche, T. Rabe, J. Stumpe, and G. Kofod, "Electrically tunable polymer DFB laser," *Advanced Materials*, vol. 23, pp. 4265–4269, 2011.

[207] C. Vannahme, C. L. C. Smith, M. B. Christiansen, and A. Kristensen, "Emission wavelength of multilayer distributed feedback dye lasers," *Applied Physics Letters*, vol. 101, p. 151123, 2012.

[208] M. Gersborg-Hansen and A. Kristensen, "Tunability of optofluidic distributed feedback dye lasers," *Optics Express*, vol. 15, pp. 137–142, 2007.

[209] F. Bernal Arango, M. B. Christiansen, M. Gersborg-Hansen, and A. Kristensen, "Optofluidic tuning of photonic crystal band edge lasers," *Applied Physics Letters*, vol. 91, p. 223503, 2007.

[210] B. Schütte, H. Gothe, S. I. Hintschich, M. Sudzius, H. Fröb, V. G. Lyssenko, and K. Leo, "Continuously tunable laser emission from a wedge-shaped organic microcavity," *Applied Physics Letters*, vol. 92, p. 163309, 2008.

[211] G. Duplain, P. G. Verly, J. A. Dobrowolski, A. Waldorf, and S. Bussière, "Graded-reflectance mirrors for beam quality control in laser resonators," *Applied Optics*, vol. 32, pp. 1145–1153, 1993.

[212] S. Klinkhammer, T. Woggon, C. Vannahme, U. Geyer, T. Mappes, and U. Lemmer, "Optical spectroscopy with organic semiconductor lasers," in *Proc. SPIE*, vol. 7722, p. 77221I, 2010.

[213] M. Stroisch, T. Woggon, C. Teiwes-Morin, S. Klinkhammer, K. Forberich, A. Gombert, M. Gerken, and U. Lemmer, "Intermediate high index layer for laser mode tuning in organic

semiconductor lasers," *Optics Express*, vol. 18, pp. 5890–5895, 2010.

[214] R. Ozaki, T. Shinpo, K. Yoshino, M. Ozaki, and H. Moritake, "Tunable liquid crystal laser using distributed feedback cavity fabricated by nanoimprint lithography," *Applied Physics Express*, vol. 1, p. 012003, 2008.

[215] S. Klinkhammer, N. Heussner, K. Huska, T. Bocksrocker, F. Geislhöringer, C. Vannahme, T. Mappes, and U. Lemmer, "Voltage-controlled tuning of an organic semiconductor distributed feedback laser using liquid crystals," *Applied Physics Letters*, vol. 99, p. 023307, 2011.

[216] A. Smekal, "Zur Quantentheorie der Dispersion," *Naturwissenschaften*, vol. 11, pp. 873–875, 1923.

[217] S. K. Freeman, *Applications of laser Raman spectroscopy*. New York: Wiley, 1974.

[218] D. A. Long, *Raman spectroscopy*. New York: McGraw-Hill, 1977.

[219] W. H. H. Weber, Ed., *Raman scattering in materials science*. Berlin: Springer, 2000.

[220] R. H. Salzer, Ed., *Infrared and Raman spectroscopic imaging*. Weinheim: Wiley-VCH, 2009.

[221] K. Kneipp, H. Kneipp, I. Itzkan, R. R. Dasari, and M. S. Feld, "Surface-enhanced Raman scattering and biophysics," *Journal of Physics: Condensed Matter*, vol. 14, pp. R597–R624, 2002.

[222] E. Smith and G. Dent, *Modern Raman spectroscopy : a practical approach*. Chichester: Wiley, 2005.

[223] S. K. Sahoo, S. Umapathy, and A. W. Parker, "Time-resolved resonance Raman spectroscopy: Exploring reactive intermediates," *Applied Spectroscopy*, vol. 65, pp. 1087–1115, 2011.

[224] J. B. Pawley, Ed., *Handbook of biological confocal microscopy*. New York: Springer, 2006.

[225] T. H. Wilson, Ed., *Confocal microscopy*. London: Academic Press, 1990.

[226] L. Novotny and B. Hecht, *Principles of nano-optics*. Cambridge: Cambridge University Press, 2007.

[227] T. Dabbs and M. Glass, "Single-mode fibers used as confocal microscope pinholes," *Applied Optics*, vol. 31, pp. 705–706, 1992.

[228] M. Fleischmann, P. Hendra, and A. McQuillan, "Raman spectra of pyridine adsorbed at a silver electrode," *Chemical Physics Letters*, vol. 26, pp. 163–166, 1974.

[229] D. L. Jeanmaire and R. P. Van Duyne, "Surface raman spectroelectrochemistry: Part I. heterocyclic, aromatic, and aliphatic amines adsorbed on the anodized silver electrode," *Journal of Electroanalytical Chemistry and Interfacial Electrochemistry*, vol. 84, pp. 1–20, 1977.

[230] M. G. Albrecht and J. A. Creighton, "Anomalously intense Raman spectra of pyridine at a silver electrode," *Journal of the American Chemical Society*, vol. 99, pp. 5215–5217, 1977.

[231] Hamamatsu Photonics K.K., "Nanophotonics," http://www.hamamatsu.com/jp/en/technology/innovation/nanophotonics/index.html, 2014.

[232] P. L. Stiles, J. A. Dieringer, N. C. Shah, and R. P. Van Duyne, "Surface-enhanced Raman spectroscopy," *Annual Review of Analytical Chemistry*, vol. 1, pp. 601–626, 2008.

[233] K. H. Kneipp, Ed., *Surface-enhanced Raman scattering: physics and applications*. Berlin: Springer, 2006.

[234] R. Aroca, R. Alvarez-Puebla, N. Pieczonka, S. Sanchez-Cortez, and J. Garcia-Ramos, "Surface-enhanced Raman scattering on colloidal nanostructures," *Advances in Colloid and Interface Science*, vol. 116, pp. 45–61, 2005.

[235] R. J. C. Brown, J. Wang, R. Tantra, R. E. Yardley, and M. J. T. Milton, "Electromagnetic modelling of Raman enhancement from

nanoscale substrates: a route to estimation of the magnitude of the chemical enhancement mechanism in SERS," *Faraday Discussions*, vol. 132, pp. 201–213, 2006.

[236] M. Moskovits, "Surface-enhanced spectroscopy," *Reviews of Modern Physics*, vol. 57, pp. 783–826, 1985.

[237] ——, "Surface selection rules," *The Journal of Chemical Physics*, vol. 77, pp. 4408–4416, 1982.

[238] J. Cabalo, J. A. Guicheteau, and S. Christesen, "Toward understanding the influence of intermolecular interactions and molecular orientation on the chemical enhancement of SERS," *The Journal of Physical Chemistry A*, vol. 117, pp. 9028–9038, 2013.

[239] J. Creighton, "The effective Raman tensor for SER scattering by molecules adsorbed at the surface of a spherical particle," *Surface Science*, vol. 158, pp. 211–221, 1985.

[240] B. Persson, "On the theory of surface-enhanced Raman scattering," *Chemical Physics Letters*, vol. 82, pp. 561–565, 1981.

[241] S. K. Saikin, Y. Chu, D. Rappoport, K. B. Crozier, and A. Aspuru-Guzik, "Separation of electromagnetic and chemical contributions to surface-enhanced Raman spectra on nanoengineered plasmonic substrates," *The Journal of Physical Chemistry Letters*, vol. 1, pp. 2740–2746, 2010.

[242] R. Pease and S. Chou, "Lithography and other patterning techniques for future electronics," *Proceedings of the IEEE*, vol. 96, pp. 248–270, 2008.

[243] N. Hendricks and K. Carter, "Nanoimprint lithography of polymers," in *Polymer Science: A Comprehensive Reference*, K. Matyjaszewski and M. Moeller, Eds. Amsterdam: Elsevier, 2012, pp. 251–274.

[244] S. Y. Chou, P. R. Krauss, and P. J. Renstrom, "Nanoimprint lithography," *Journal of Vacuum Science & Technology B*, vol. 14, pp. 4129–4133, 1996.

[245] W. Menz, J. Mohr, and O. Paul, *Microsystem technology*. Weinheim: Wiley-VCH, 2001.

[246] L. Guo, “Nanoimprint lithography: Methods and material requirements,” *Advanced Materials*, vol. 19, pp. 495–513, 2007.

[247] C. Vannahme, “Integration organischer Laser in Lab-on-Chip Systeme,” Ph.D. dissertation, Karlsruhe Institute of Technology, Karlsruhe, 2011.

[248] V. Grigaliunas, S. Tamulevicius, R. Tomasiunas, V. Kopustinskas, A. Guobien, and D. Jucius, “Laser pulse assisted nanoimprint lithography,” *Thin Solid Films*, vol. 453-454, pp. 13–15, 2004.

[249] V. Grigaliunas, S. Tamulevicius, M. Muehlberger, D. Jucius, A. Guobiene, V. Kopustinskas, and A. Gudonyte, “Nanoimprint lithography using IR laser irradiation,” *Applied Surface Science*, vol. 253, pp. 646–650, 2006.

[250] W. Pfleging, R. Kohler, I. Südmeyer, and M. Rohde, “Laser micro and nano processing of metals, ceramics, and polymers,” in *Springer Series in Materials Science*, J. D. Majumdar and I. Manna, Eds. Springer, 2013, vol. 161, pp. 319–374.

[251] M. Wissmann, H. Besser, M. Beiser, and W. Pfleging, “Laser moulding, a new low-cost fabrication process for micro- and nanostructured components,” *Microsystem Technologies*, pp. 1–7, 2014.

[252] S. Prinz, “Herstellung von funktionellen Nanostrukturen durch laserunterstütztes Heißprägen,” Master’s thesis, Karlruhe Institute of Technology, Karlsruhe, 2013.

[253] M. J. Beesley and J. G. Castledine, “The use of photoresist as a holographic recording medium,” *Applied Optics*, vol. 9, pp. 2720–2724, 1970.

[254] G. Guzman, B. Dahmani, J. Puetz, and M. Aegerter, “Transparent conducting sol-gel ATO coatings for display applications by an improved dip coating technique,” *Thin Solid Films*, vol. 502, pp. 281–285, 2006.

[255] A. L. Briseno, M. Roberts, M.-M. Ling, H. Moon, E. J. Nemanick, and Z. Bao, "Patterning organic semiconductors using "dry" poly(dimethylsiloxane) elastomeric stamps for thin film transistors," *Journal of the American Chemical Society*, vol. 128, pp. 3880–3881, 2006.

[256] I.-K. Ding, J. Melas-Kyriazi, N.-L. Cevey-Ha, K. G. Chittibabu, S. M. Zakeeruddin, M. Grätzel, and M. D. McGehee, "Deposition of hole-transport materials in solid-state dye-sensitized solar cells by doctor-blading," *Organic Electronics*, vol. 11, pp. 1217–1222, 2010.

[257] G. Jabbour, R. Radspinner, and N. Peyghambarian, "Screen printing for the fabrication of organic light-emitting devices," *Journal of Selected Topics in Quantum Electronics*, vol. 7, pp. 769–773, 2001.

[258] S. E. Shaheen, R. Radspinner, N. Peyghambarian, and G. E. Jabbour, "Fabrication of bulk heterojunction plastic solar cells by screen printing," *Applied Physics Letters*, vol. 79, pp. 2996–2998, 2001.

[259] T. R. Hebner, C. C. Wu, D. Marcy, M. H. Lu, and J. C. Sturm, "Ink-jet printing of doped polymers for organic light emitting devices," *Applied Physics Letters*, vol. 72, pp. 519–521, 1998.

[260] H. Sirringhaus, T. Kawase, R. H. Friend, T. Shimoda, M. Inbasekaran, W. Wu, and E. P. Woo, "High-resolution inkjet printing of all-polymer transistor circuits," *Science*, vol. 290, pp. 2123–2126, 2000.

[261] C. N. Hoth, P. Schilinsky, S. A. Choulis, and C. J. Brabec, "Printing highly efficient organic solar cells," *Nano Letters*, vol. 8, pp. 2806–2813, 2008.

[262] B. Park and M.-y. Han, "Photovoltaic characteristics of polymer solar cells fabricated by pre-metered coating process," *Optics Express*, vol. 17, pp. 13 830–13 840, 2009.

[263] C. Ge, M. Lu, X. Jian, Y. Tan, and B. T. Cunningham, "Large-area organic distributed feedback laser fabricated by nanoreplica molding and horizontal dipping," *Optics Express*, vol. 18, pp. 12 980–12 991, 2010.

[264] B. Park and M.-y. Han, "Organic light-emitting devices fabricated using a premetered coating process," *Optics Express*, vol. 17, pp. 21 362–21 369, 2009.

[265] L. D. Landau and V. G. Levich, "Dragging of a liquid by a moving plate," *Acta Physicochimica URSS*, vol. 17, pp. 42–54, 1942.

[266] X. Liu, "Solution processing and characterization of organic DFB lasers," Master's thesis, Karlruhe Institute of Technology, Karlsruhe, 2011.

[267] F. Nickel, C. Sprau, M. F. Klein, P. Kapetana, N. Christ, X. Liu, S. Klinkhammer, U. Lemmer, and A. Colsmann, "Spatial mapping of photocurrents in organic solar cells comprising wedge-shaped absorber layers for an efficient material screening," *Solar Energy Materials and Solar Cells*, vol. 104, pp. 18–22, 2012.

[268] S. Höfle, T. Lutz, A. Egel, F. Nickel, S. W. Kettlitz, G. Gomard, U. Lemmer, and A. Colsmann, "Influence of the emission layer thickness on the optoelectronic properties of solution processed organic light-emitting diodes," *ACS Photonics*, vol. 1, pp. 968–973, 2014.

[269] W. Plass, R. Maestle, K. Wittig, A. Voss, and A. Giesen, "High-resolution knife-edge laser beam profiling," *Optics Communications*, vol. 134, pp. 21–24, 1997.

[270] X. Liu, S. Klinkhammer, K. Sudau, N. Mechau, C. Vannahme, J. Kaschke, T. Mappes, M. Wegener, and U. Lemmer, "Ink-jet-printed organic semiconductor distributed feedback laser," *Applied Physics Express*, vol. 5, p. 072101, 2012.

[271] X. Liu, S. Prinz, H. Besser, W. Pfleging, M. Wissmann, C. Vannahme, M. Guttmann, T. Mappes, S. Koeber, C. Koos, and U. Lemmer, "Organic semiconductor distributed feedback laser pixels for lab-on-a-chip applications fabricated by laser-assisted replication," *Faraday Discussions*, vol. 174, pp. 153–164, 2014.

[272] O. Geschke, Ed., *Microsystem engineering of lab on a chip devices*. Weinheim: Wiley-VCH, 2008.

[273] R. Daw and J. Finkelstein, "Lab on a chip," *Nature*, vol. 442, p. 367, 2006.

[274] C. Vannahme, S. Klinkhammer, M. B. Christiansen, A. Kolew, A. Kristensen, U. Lemmer, and T. Mappes, "All-polymer organic semiconductor laser chips: Parallel fabrication and encapsulation," *Optics Express*, vol. 18, pp. 24 881–24 887, 2010.

[275] X. Liu, S. Klinkhammer, Z. Wang, T. Wienhold, C. Vannahme, P.-J. Jakobs, A. Bacher, A. Muslija, T. Mappes, and U. Lemmer, "Pump spot size dependent lasing threshold in organic semiconductor DFB lasers fabricated via nanograting transfer," *Optics Express*, vol. 21, pp. 27 697–27 706, 2013.

[276] R. Stabile, A. Camposeo, L. Persano, S. Tavazzi, R. Cingolani, and D. Pisignano, "Organic-based distributed feedback lasers by direct electron-beam lithography on conjugated polymers," *Applied Physics Letters*, vol. 91, p. 101110, 2007.

[277] M. Stroisch, T. Woggon, U. Lemmer, G. Bastian, G. Violakis, and S. Pissadakis, "Organic semiconductor distributed feedback laser fabricated by direct laser interference ablation," *Optics Express*, vol. 15, pp. 3968–3973, 2007.

[278] H. E. Katz, "Recent advances in semiconductor performance and printing processes for organic transistor-based electronics," *Chemistry of Materials*, vol. 16, pp. 4748–4756, 2004.

[279] J. Bharathan and Y. Yang, "Polymer electroluminescent devices processed by inkjet printing: I. Polymer light-emitting logo," *Applied Physics Letters*, vol. 72, pp. 2660–2662, 1998.

[280] T. Aernouts, T. Aleksandrov, C. Girotto, J. Genoe, and J. Poortmans, "Polymer based organic solar cells using ink-jet printed active layers," *Applied Physics Letters*, vol. 92, p. 033306, 2008.

[281] E. Tekin, E. Holder, D. Kozodaev, and U. Schubert, "Controlled pattern formation of poly[2-methoxy-5-(2'-ethylhexyloxyl)-1,4-phenylenevinylene] (MEH-PPV) by ink-jet printing," *Advanced Functional Materials*, vol. 17, pp. 277–284, 2007.

[282] B.-J. de Gans and U. S. Schubert, “Inkjet printing of well-defined polymer dots and arrays,” *Langmuir*, vol. 20, pp. 7789–7793, 2004.

[283] R. Pal and E. Rhodes, “Viscosity/concentration relationships for emulsions,” *Journal of Rheology*, vol. 33, pp. 1021–1045, 1989.

[284] R. Ober, L. Paz, C. Taupin, P. Pincus, and S. Boileau, “Study of the surface tension of polymer solutions: theory and experiments. good solvent conditions,” *Macromolecules*, vol. 16, pp. 50–55, 1983.

[285] S. M. Jeong, N. Y. Ha, F. Araoka, K. Ishikawa, and H. Takezoe, “Electrotunable polarization of surface-emitting distributed feedback laser with nematic liquid crystals,” *Applied Physics Letters*, vol. 92, p. 171105, 2008.

[286] X. Cheng and L. Jay Guo, “One-step lithography for various size patterns with a hybrid mask-mold,” *Microelectronic Engineering*, vol. 71, pp. 288–293, 2004.

[287] C. Vannahme, M. B. Christiansen, T. Mappes, and A. Kristensen, “Optofluidic dye laser in a foil,” *Optics Express*, vol. 18, pp. 9280–9285, 2010.

[288] W. Pfleging, O. Baldus, and A. Baldini, “Method for joining plastic workpieces,” European Patent No. EP20 050 739 878, November, 2013.

[289] X. Liu, S. Lebedkin, T. Mappes, S. Köber, C. Koos, M. Kappes, and U. Lemmer, “Organic semiconductor distributed feedback laser as excitation source in Raman spectroscopy using free-beam and fibre coupling,” in *Proc. SPIE*, vol. 9137, p. 91370Y, 2014.

[290] A. D. McFarland, M. A. Young, J. A. Dieringer, and R. P. Van Duyne, “Wavelength-scanned surface-enhanced raman excitation spectroscopy,” *J. Phys. Chem. B*, vol. 109, pp. 11 279–11 285, 2005.

[291] E. M. Calzado, J. M. Villalvilla, P. G. Boj, J. A. Quintana, V. Navarro-Fuster, A. Retolaza, S. Merino, and M. A. Díaz-García,

"Influence of the excitation area on the thresholds of organic second-order distributed feedback lasers," *Applied Physics Letters*, vol. 101, p. 223303, 2012.

[292] B. K. Yap, R. Xia, M. Campoy-Quiles, P. N. Stavrinou, and D. D. C. Bradley, "Simultaneous optimization of charge-carrier mobility and optical gain in semiconducting polymer films," *Nature Materials*, vol. 7, pp. 376–380, 2008.

[293] E. B. Namdas, M. Tong, P. Ledochowitsch, S. R. Mednick, J. D. Yuen, D. Moses, and A. J. Heeger, "Low thresholds in polymer lasers on conductive substrates by distributed feedback nanoimprinting: Progress toward electrically pumped plastic lasers," *Advanced Materials*, vol. 21, pp. 799–802, 2009.

[294] N. Tsutsumi and H. Nishida, "Tunable distributed feedback lasing with low threshold and high slope efficiency from electroluminescent conjugated polymer waveguide," *Optics Communications*, vol. 284, pp. 3365–3368, 2011.

[295] C. Grivas and M. Pollnau, "Organic solid-state integrated amplifiers and lasers," *Laser & Photonics Reviews*, vol. 6, pp. 419–462, 2012.

[296] E. R. Martins, Y. Wang, A. L. Kanibolotsky, P. J. Skabara, G. A. Turnbull, and I. D. W. Samuel, "Low-threshold nanoimprinted lasers using substructured gratings for control of distributed feedback," *Advanced Optical Materials*, vol. 1, pp. 563–566, 2013.

[297] V. Navarro-Fuster, I. Vragovic, E. M. Calzado, P. G. Boj, J. A. Quintana, J. M. Villalvilla, A. Retolaza, A. Juarros, D. Otaduy, S. Merino, and M. A. Díaz-García, "Film thickness and grating depth variation in organic second-order distributed feedback lasers," *Journal of Applied Physics*, vol. 112, p. 043104, 2012.

[298] S. Döring, T. Rabe, and J. Stumpe, "Output characteristics of organic distributed feedback lasers with varying grating heights," *Applied Physics Letters*, vol. 104, p. 263302, 2014.

[299] N. Tsutsumi and M. Shinobu, "All organic DFB laser enhanced by intermediate high refractive index polymer layer," *Applied Physics B*, vol. 105, pp. 839–845, 2011.

[300] W. Zhao, T. Cao, and J. White, "On the origin of green emission in polyfluorene polymers: The roles of thermal oxidation degradation and crosslinking," *Advanced Functional Materials*, vol. 14, pp. 783–790, 2004.

[301] A. Costela, I. García-Moreno, D. del Agua, O. García, and R. Sastre, "Silicon-containing organic matrices as hosts for highly photostable solid-state dye lasers," *Applied Physics Letters*, vol. 85, pp. 2160–2162, 2004.

[302] S. Scholz, C. Corten, K. Walzer, D. Kuckling, and K. Leo, "Photochemical reactions in organic semiconductor thin films," *Organic Electronics*, vol. 8, pp. 709–717, 2007.

[303] T. L. Andrew and T. M. Swager, "Reduced photobleaching of conjugated polymer films through small molecule additives," *Macromolecules*, vol. 41, pp. 8306–8308, 2008.

[304] S. Scholz, R. Meerheim, K. Walzer, and K. Leo, "Chemical degradation mechanisms of organic semiconductor devices," in *Proc. SPIE*, vol. 6999, p. 69991B, 2008.

[305] F. So and D. Kondakov, "Degradation mechanisms in small-molecule and polymer organic light-emitting diodes," *Advanced Materials*, vol. 22, pp. 3762–3777, 2010.

[306] S. Richardson, O. P. M. Gaudin, G. A. Turnbull, and I. D. W. Samuel, "Improved operational lifetime of semiconducting polymer lasers by encapsulation," *Applied Physics Letters*, vol. 91, p. 261104, 2007.

[307] L. Persano, A. Camposeo, P. Del Carro, P. Solaro, R. Cingolani, P. Boffi, and D. Pisignano, "Rapid prototyping encapsulation for polymer light-emitting lasers," *Applied Physics Letters*, vol. 94, p. 123305, 2009.

[308] X. Wang, X. Gao, Y. Zhou, Z. Xie, Q. Song, X. Ding, and X. Hou, "Photodegradation of organic light-emitting devices observed in nitrogen-filled environment," *Thin Solid Films*, vol. 516, pp. 2171–2174, 2008.

[309] Schott AG, "Schott D263®," http://www.schott.com/advanced_optics/german/syn/advanced_optics/products/optical-materials/thin-glass/cover-glass-d-263-m/index.html, 2014.

[310] TOPAS Advanced Polymers, Inc., "TOPAS® 6013," http://www.topas.com/sites/default/files/files/optical_e.pdf, 2011.

[311] MicroChem Corp., "SU-8 2000," http://www.microchem.com/pdf/SU-82000DataSheet2000_5thru2015Ver4.pdf, 2014.

[312] Evonik Industries AG, "Plexiglas® XT," http://www.plexiglas.net/sites/dc/Downloadcenter/Evonik/Product/PLEXIGLAS-Sheet/PLEXIGLAS/232-6-PLEXIGLAS-XT-UV-durchlaessig-de.pdf, 2011.

[313] Allresist GmbH, "SX AR-N 4800/16," http://www.allresist.com/wp-content/uploads/sites/2/wMedia/pdf/wEnglish/produkte_sonder/SXAR-N4800_16.pdf, 2007.

[314] E. D. Palik, Ed., *Handbook of optical constants of solids*. New York: Academic Press, 1991.

[315] A. T. Ward, "Raman spectroscopy of sulfur, sulfur-selenium, and sulfur-arsenic mixtures," *The Journal of Physical Chemistry*, vol. 72, pp. 4133–4139, 1968.

[316] S. N. White, R. M. Dunk, E. T. Peltzer, J. J. Freeman, and P. G. Brewer, "In situ Raman analyses of deep-sea hydrothermal and cold seep systems (gorda ridge and hydrate ridge)," *Geochemistry, Geophysics, Geosystems*, vol. 7, p. Q05023, 2006.

[317] S. N. White, "Laser Raman spectroscopy as a technique for identification of seafloor hydrothermal and cold seep minerals," *Chemical Geology*, vol. 259, pp. 240–252, 2009.

[318] D. Himmel, L. C. Maurin, O. Gros, and J.-L. Mansot, "Raman microspectrometry sulfur detection and characterization in the marine ectosymbiotic nematode eubostrichus dianae (desmodoridae, stilbonematidae)," *Biology of the Cell*, vol. 101, pp. 43–54, 2009.

[319] R. T. Downs, "The RRUFF project: an integrated study of the chemistry, crystallography, Raman and infrared spectroscopy of minerals." in *Program and Abstracts of the 19th General Meeting of the International Mineralogical Association*, no. O03-13, Kobe, Japan, 2006.

[320] C. A. Arguello, D. L. Rousseau, and S. P. S. Porto, "First-order Raman effect in wurtzite-type crystals," *Physical Review*, vol. 181, pp. 1351–1363, 1969.

[321] M. Abdulkhadar and B. Thomas, "Study of Raman spectra of nanoparticles of CdS and ZnS," *Nanostructured Materials*, vol. 5, pp. 289–298, 1995.

[322] F. Gu, H. Yu, W. Fang, and L. Tong, "Low-threshold supercontinuum generation in semiconductor nanoribbons by continuous-wave pumping," *Optics Express*, vol. 20, pp. 8667–8674, 2012.

[323] X. Liu, S. Lebedkin, H. Besser, W. Pfleging, S. Prinz, M. Wissmann, P. M. Schwab, I. Nazarenko, M. Guttmann, M. M. Kappes, and U. Lemmer, "Tailored surface-enhanced raman nanopillar arrays fabricated by laser-assisted replication for biomolecular detection using organic semiconductor lasers," *ACS Nano*, vol. 9, pp. 260–270, 2015.

[324] I. Nabiev, I. Chourpa, and M. Manfait, "Applications of Raman and surface-enhanced Raman scattering spectroscopy in medicine," *Journal of Raman Spectroscopy*, vol. 25, pp. 13–23, 1994.

[325] L.-P. Choo-Smith, H. G. M. Edwards, H. P. Endtz, J. M. Kros, F. Heule, H. Barr, J. S. Robinson, H. A. Bruining, and G. J. Puppels, "Medical applications of Raman spectroscopy: From proof of principle to clinical implementation," *Biopolymers*, vol. 67, pp. 1–9, 2002.

[326] J. S. Mak, S. A. Rutledge, R. M. Abu-Ghazalah, F. Eftekhari, J. Irizar, N. C. Tam, G. Zheng, and A. S. Helmy, "Recent developments in optofluidic-assisted Raman spectroscopy," *Progress in Quantum Electronics*, vol. 37, pp. 1–50, 2013.

[327] W. F. Paxton, S. L. Kleinman, A. N. Basuray, J. F. Stoddart, and R. P. Van Duyne, "Surface-enhanced Raman spectroelectrochemistry of TTF-modified self-assembled monolayers," *The Journal of Physical Chemistry Letters*, vol. 2, pp. 1145–1149, 2011.

[328] F. Adar, R. Geiger, and J. Noonan, "Raman spectroscopy for process/quality control," *Applied Spectroscopy Reviews*, vol. 32, pp. 45–101, 1997.

[329] C. L. Brosseau, K. S. Rayner, F. Casadio, C. M. Grzywacz, and R. P. Van Duyne, "Surface-enhanced Raman spectroscopy: A direct method to identify colorants in various artist media," *Analytical Chemistry*, vol. 81, pp. 7443–7447, 2009.

[330] B. Sharma, R. R. Frontiera, A.-I. Henry, E. Ringe, and R. P. Van Duyne, "SERS: Materials, applications, and the future," *Materials Today*, vol. 15, pp. 16–25, 2012.

[331] S. Dochow, C. Beleites, T. Henkel, G. Mayer, J. Albert, J. Clement, C. Krafft, and J. Popp, "Quartz microfluidic chip for tumour cell identification by Raman spectroscopy in combination with optical traps," *Analytical and Bioanalytical Chemistry*, vol. 405, pp. 2743–2746, 2013.

[332] S. Dochow, M. Becker, R. Spittel, C. Beleites, S. Stanca, I. Latka, K. Schuster, J. Kobelke, S. Unger, T. Henkel, G. Mayer, J. Albert, M. Rothhardt, C. Krafft, and J. Popp, "Raman-on-chip device and detection fibres with fibre Bragg grating for analysis of solutions and particles," *Lab on a Chip*, vol. 13, pp. 1109–1113, 2013.

[333] G. L. Liu and L. P. Lee, "Nanowell surface enhanced Raman scattering arrays fabricated by soft-lithography for label-free biomolecular detections in integrated microfluidics," *Applied Physics Letters*, vol. 87, p. 074101, 2005.

[334] C. Lim, J. Hong, B. G. Chung, A. J. deMello, and J. Choo, "Optofluidic platforms based on surface-enhanced Raman scattering," *Analyst*, vol. 135, pp. 837–844, 2010.

[335] W. Fritzsche and P. Jürgen, Eds., *Optical nano- and microsystems for bioanalytics*. Berlin: Springer, 2012.

[336] S. Lee, J. Choi, L. Chen, B. Park, J. B. Kyong, G. H. Seong, J. Choo, Y. Lee, K.-H. Shin, E. K. Lee, S.-W. Joo, and K.-H. Lee, "Fast and sensitive trace analysis of malachite green using a surface-enhanced Raman microfluidic sensor," *Analytica Chimica Acta*, vol. 590, pp. 139–144, 2007.

[337] L. X. Quang, C. Lim, G. H. Seong, J. Choo, K. J. Do, and S.-K. Yoo, "A portable surface-enhanced Raman scattering sensor integrated with a lab-on-a-chip for field analysis," *Lab on a Chip*, vol. 8, pp. 2214–2219, 2008.

[338] M. Wang, N. Jing, I.-H. Chou, G. L. Cote, and J. Kameoka, "An optofluidic device for surface enhanced Raman spectroscopy," *Lab on a Chip*, vol. 7, pp. 630–632, 2007.

[339] L. Tong, M. Righini, M. U. Gonzalez, R. Quidant, and M. Kall, "Optical aggregation of metal nanoparticles in a microfluidic channel for surface-enhanced Raman scattering analysis," *Lab on a Chip*, vol. 9, pp. 193–195, 2009.

[340] R. M. Connatser, M. Cochran, R. J. Harrison, and M. J. Sepaniak, "Analytical optimization of nanocomposite surface-enhanced Raman spectroscopy/scattering detection in microfluidic separation devices," *Electrophoresis*, vol. 29, pp. 1441–1450, 2008.

[341] A. Banerjee, R. Perez-Castillejos, D. Hahn, A. I. Smirnov, and H. Grebel, "Micro-fluidic channels on nanopatterned substrates: Monitoring protein binding to lipid bilayers with surface-enhanced Raman spectroscopy," *Chemical Physics Letters*, vol. 489, pp. 121–126, 2010.

[342] R. Tantra, R. J. C. Brown, and M. J. T. Milton, "Strategy to improve the reproducibility of colloidal SERS," *Journal of Raman Spectroscopy*, vol. 38, pp. 1469–1479, 2007.

[343] L. Maya, C. E. Vallet, and Y. H. Lee, "Sputtered gold films for surface-enhanced Raman scattering," *Journal of Vacuum Science & Technology A*, vol. 15, pp. 238–242, 1997.

[344] K. L. Norrod, L. M. Sudnik, D. Rousell, and K. L. Rowlen, "Quantitative comparison of five SERS substrates: Sensitivity and limit of detection," *Applied Spectroscopy*, vol. 51, pp. 994–1001, 1997.

[345] T. Zhu, H. Yu, J. Wang, Y. Wang, S. Cai, and Z. Liu, "Two-dimensional surface enhanced Raman mapping of differently prepared gold substrates with an azobenzene self-assembled monolayer," *Chemical Physics Letters*, vol. 265, pp. 334–340, 1997.

[346] L.-X. Xia, K. Gong, J. Wang, X.-B. Kang, S.-R. Tong, and G.-Y. Liu, "A simple method to fabricate patterned 2-dimensional silver nano-arrays," *Acta Chimica Sinica*, vol. 65, pp. 2489–2493, 2007.

[347] J. Fang, S. Du, S. Lebedkin, Z. Li, R. Kruk, M. Kappes, and H. Hahn, "Gold mesostructures with tailored surface topography and their self-assembly arrays for surface-enhanced Raman spectroscopy," *Nano Letters*, vol. 10, pp. 5006–5013, 2010.

[348] M. Green and F. M. Liu, "SERS substrates fabricated by island lithography: The silver/pyridine system," *Journal of Materials Chemistry B*, vol. 107, pp. 13 015–13 021, 2003.

[349] C. J. Choi, Z. Xu, H.-Y. Wu, G. L. Liu, and B. T. Cunningham, "Surface-enhanced Raman nanodomes," *Nanotechnology*, vol. 21, p. 415301, 2010.

[350] J.-A. Huang, Y.-Q. Zhao, X.-J. Zhang, L.-F. He, T.-L. Wong, Y.-S. Chui, W.-J. Zhang, and S.-T. Lee, "Ordered Ag/Si nanowires array: Wide-range surface-enhanced Raman spectroscopy for reproducible biomolecule detection," *Nano Letters*, vol. 13, pp. 5039–5045, 2013.

[351] J. C. Hulteen and R. P. Van Duyne, "Nanosphere lithography: A materials general fabrication process for periodic particle array surfaces," *Journal of Vacuum Science & Technology A*, vol. 13, pp. 1553–1558, 1995.

[352] S. Goyal, M. R. Thorson, C. L. Schneider, G. G. Z. Zhang, Y. Gong, and P. J. A. Kenis, "A microfluidic platform for evaporation-based salt screening of pharmaceutical parent compounds," *Lab on a Chip*, vol. 13, pp. 1708–1723, 2013.

[353] F. Pallikari, G. Chondrokoukis, M. Rebelakis, and Y. Kotsalas, "Raman spectroscopy: A technique for estimating extent of polymerization in pmma," *Material Research Innovations*, vol. 4, pp. 89–92, 2001.

[354] K. Kneipp, "Surface-enhanced Raman scattering," *Physics Today*, vol. 60, pp. 40–46, 2007.

[355] M. J. Mulvihill, X. Y. Ling, J. Henzie, and P. Yang, "Anisotropic etching of silver nanoparticles for plasmonic structures capable of single-particle SERS," *Journal of the American Chemical Society*, vol. 132, pp. 268–274, 2009.

[356] M. D. Wissert, A. W. Schell, K. S. Ilin, M. Siegel, and H.-J. Eisler, "Nanoengineering and characterization of gold dipole nanoantennas with enhanced integrated scattering properties," *Nanotechnology*, vol. 20, p. 425203, 2009.

[357] P. P. Yaney, "The pulsed laser and gated detection in Raman spectroscopy - a survey of the spectra of common substances including studies of adsorbed benzene," *Journal of Raman Spectroscopy*, vol. 5, pp. 219–241, 1976.

[358] P. Matousek, M. Towrie, A. Stanley, and A. W. Parker, "Efficient rejection of fluorescence from Raman spectra using picosecond kerr gating," *Applied Spectroscopy*, vol. 53, pp. 1485–1489, 1999.

[359] F. Knorr, Z. J. Smith, and S. Wachsmann-Hogiu, "Development of a time-gated system for Raman spectroscopy of biological samples," *Optics Express*, vol. 18, pp. 20 049–20 058, 2010.

[360] I. Nissinen, J. Nissinen, A. Lansman, L. Hallman, A. Kilpela, J. Kostamovaara, M. Kogler, M. Aikio, and J. Tenhunen, "A sub-ns time-gated CMOS single photon avalanche diode detector for Raman spectroscopy," in *Solid-State Device Research Conference (ESSDERC), 2011 Proceedings of the European*, pp. 375–378, 2011.

[361] J. Kostamovaara, J. Tenhunen, M. Kögler, I. Nissinen, J. Nissinen, and P. Keränen, "Fluorescence suppression in Raman spectroscopy

using a time-gated CMOS SPAD," *Optics Express*, vol. 21, pp. 31 632–31 645, 2013.

[362] M. M. Carrabba, K. M. Spencer, C. Rich, and D. Rauh, "The utilization of a holographic Bragg diffraction filter for rayleigh line rejection in Raman spectroscopy," *Applied Spectroscopy*, vol. 44, pp. 1558–1561, 1990.

[363] C. L. Schoen, S. K. Sharma, C. E. Helsley, and H. Owen, "Performance of a holographic supernotch filter," *Applied Spectroscopy*, vol. 47, pp. 305–308, 1993.

[364] M. Paillet, F. Meunier, M. Verhaegen, S. Blais-Ouellette, and R. Martel, "High performance resonance Raman spectroscopy using volume Bragg gratings as tunable light filters," *Review of Scientific Instruments*, vol. 81, p. 053111, 2010.

[365] N. Stürzl, S. Lebedkin, S. Klumpp, F. Hennrich, and M. M. Kappes, "Novel micro-Raman setup with tunable laser excitation for time-efficient resonance Raman microscopy and imaging," *Analytical Chemistry*, vol. 85, pp. 4554–4559, 2013.

[366] K. W. Kho, K. Z. M. Qing, Z. X. Shen, I. B. Ahmad, S. S. C. Lim, S. Mhaisalkar, T. J. White, F. Watt, K. C. Soo, and M. Olivo, "Polymer-based microfluidics with surface-enhanced Raman-spectroscopy-active periodic metal nanostructures for biofluid analysis," *Journal of Biomedical Optics*, vol. 13, p. 054026, 2008.

Appendix A: Recipes

A.1 Fabrication of silicon template

Electron beam lithography

Process	Description	Duration
Labeling	n-doped Silicon wafer	
Prebake	Hotplate 180 °C	10 min
Spin-coating	PMMA 950k A2 1250 rpm 150 rpm ramp	 60 s (50 nm)
Softbake	Hotplate 180 °C	5 min
E-beam exposure	Vistec VB6 Dose variation 650–800 $\mu C\ cm^{-2}$	
Development	MIBK:IPA 1:3	10 s

UV-lithography

Process	Description	Duration
Adhesion promotion	IMTEC Star 2000 HMDS deposition 140 °C 5.7 mbar	30 min
Spin-coating	AZ1505 3000 rpm	90 s (0.5 µm)
Softbake	Hotplate 110 °C	1 min
UV-exposure	EVG 620 mask aligner soft contact mode 50 mJ cm^{-2}	
Development	AZ400K:water 1:4	1 min

Reactive-ion etching

Process	Description	Duration
Precooling	−125 °C	4 min
Etching phase 1	Plasmalab 100 ICP 380 Oxford Instruments 24 sccm SF_6, 6 sccm O_2, 4.5 mTorr, 10 Torr Ar, 10 Torr He, 15 W RF, 600 W ICP, −125 °C	3 s
Etching phase 2	24 sccm SF_6, 6 sccm O_2, 4.5 mTorr, 10 Torr Ar, 10 Torr He, 8 W RF, 600 W ICP, −125 °C	3 s[1] (∼ 100 nm) /100 s[2] (∼ 1.5 µm)

[1]For fabrication of DFB gratings and SERS nanopillar arrays after EBL.
[2]For fabrication of grating trenches and waveguides after UV-lithography.

A.2 Photoresist gratings fabricated by interference lithography

SX AR-N 4800/16 grating

Process	Description	Duration
Cleaning	Ultrasound bath acetone/IPA	 5 min/5 min
Prebake	Hotplate 150 °C	10 min
Adhesion promotion	HMDS deposition	10 min
Spin-coating	SX AR-N 4800/16 500 rpm 1000 rpm	 5 s 45 s
Softbake	Hotplate 85 °C	5 min
Laser-exposure	30–50 mJ cm^{-2}	
Exposure delay		4 min
Postbake	Hotplate 80 °C	2 min
Development	X AR 600-54/3 stopped by nitrogen blow	40–70 s

SU-8 2005 grating

Process	Description	Duration
Cleaning	Ultrasound bath acetone/IPA	 5 min/5 min
Prebake	Hotplate 150 °C	10 min
Adhesion promotion	HMDS deposition	10 min
Spin-coating	SU-8 2005 500 rpm 4000 rpm	 5 s 30 s
Softbake 1	Hotplate 65 °C	2 min
Softbake 2	Hotplate 96 °C	5 min
Chilling		2 min
Laser-exposure	10–18 mJ cm^{-2}	
Exposure delay		4 min
Postbake 1	Hotplate 65 °C	1 min
Postbake 2	Hotplate 96 °C	30 s
Chilling		2 min
Development	DEV 600 stopped by nitrogen blow	30 s

A.3 E-beam doses applied for nanowell patterns

1. Nanowell spacing 20 nm: 9 lines of nanowell arrays with diameters of 50, 70, 90, 110, 130, 150, 170, 200 and 250 nm. E-beam doses for 7 columns: 400, 500, 600, 700, 800, 900 and 1000 μC cm^{-2}, respectively.

2. Spacing 200 nm: 9 lines of nanowell arrays with diameters of 50, 70, 90, 110, 130, 150, 170, 200 and 250 nm. E-beam doses for 7 columns: 500, 600, 700, 800, 900, 1000, and 1100 μC cm^{-2}, respectively.

3. Spacing equals the nanowell diameter plus 20 nm: 9 lines of nanowell arrays with diameters of 50, 70, 90, 110, 130, 150, 170, 190 and 210 nm. E-beam doses for 7 columns: 500, 600, 700, 800, 900, 1000, and 1100 μC cm^{-2}, respectively.

Appendix B: Components of Laser-Assisted Replication Apparatus

Description	Manufacturer	Type
Laser system		
Laser head	Fisba Optik AG	FLS iron 50/940 LH-018
Pyrometer	Fisba Optik AG	PyroS
Scanner system	Fisba Optik AG	FLS ironScan
Objective	Fisba Optik AG	FT-163
Sample stage	IAM-AWP	
External system		
PC control board	Scanlab AG	RTC3
Laser control unit	Fisba Optik AG	PMD-M
Power supply	Fisba Optik AG	IRON PSIF
Cooling system	Thermotek AG	P302
Software		
Data collection and temperature control	Dr. Mergenthaler GmbH	LASCON v3.74
Scanner software	Scaps GmbH	SAMLight v2.2
Laser control software	Fisba Optik AG	FisbaCom v1.2.0d

Appendix C: Abbreviations

Material	Name or IUPAC nomenclature
Ag	silver
Au	gold
Alq_3	tris-(8-hydroxyquinoline)aluminum
CdS	cadmium sulfide
COC	cyclic olefin copolymer
DCM	4-(Dicyanomethylene)-2-methyl-6-(p-dimethylaminostyryl)-4H-pyran
DNT	2,4-dinitrotoluene
F8BT	poly[9,9-dioctylfluorenyl-2,7-diyl)-co-1,4-benzo-{2,1-3}-thiadiazole)]
FDTS	perfluorodecyltrichlorosilane
GaN	gallium nitride
HfO_2	hafnium oxide
HMDS	hexamethyldisilazane
InGaN	indium gallium nitride
IPA	isopropyl alcohol
MEH-PPV	poly(2-methoxy-5-(2'-ethylhexyloxy)-1,4-phenylene vinylene)

MIBK	methyl isobutyl ketone
Nd:YAG	neodymium-doped yttrium aluminum garnet
Nd:YVO_4	neodymium-doped yttrium orthovanadate
PDMS	polydimethylsiloxane
PMMA	poly(methyl methacrylate)
PPV	poly(p-phenylen-vinylen)
S_8	sulfur
SiO_2	silicon dioxide
THF	tetrahydrofuran
TNT	2,4,6-trinitrotoluene

Acronym	**Explanation**
AFM	atomic force microscope/microscopy
ASE	amplified spontaneous emission
CARS	coherent anti-Stokes Raman scattering
CCD	charge-coupled device
CMOS	complementary metal-oxide-semiconductor
CVD	chemical vapor deposition
CW	continuous wave
DBR	distributed Bragg reflector
DFB	distributed feedback

DFB-OSL	organic semiconductor distributed feedback laser
DPSS	diode-pumped solid-state
DUV	deep ultraviolet
EBL	electron beam lithography
EF	enhancement factor
EMCCD	electron multiplying charge-coupled device
E-Beam	electron beam
FDTD	finite-difference time-domain
FRET	Förster resonant energy transfer
FWHM	full width at half maximum
HOMO	highest occupied molecular orbit
ICCD	intensified charge-coupled device
IMT	Institute of Microstructure Technology
INT	Institute of Nanotechnology
ISC	intersystem crossing
IL	interference lithography
KIT	Karlsruhe Institute of Technology
LASER	light amplification by stimulated emission of radiation
LC	liquid crystal
LED	light-emitting diode
LOC	lab-on-a-chip

LSPR	localized surface plasmon resonance
LTI	Light Technology Institute
LUMO	lowest unoccupied molecular orbit
NA	numerical aperture
NIL	nanoimprint lithography
OFET	organic field-effect transistor
OLED	organic light-emitting diode
OSL	organic semiconductor laser
OTFT	organic thin-film transistor
PLQE	photoluminescence quantum efficiency
RRS	resonance Raman spectroscopy
SEM	scanning electron microscopy
SERS	surface-enhanced Raman scattering/spectroscopy
SPAD	single-photon avalanche diode
SSA	singlet-singlet annihilation
STA	singlet-triplet annihilation
TE	transverse electric
TERS	tip-enhanced Raman spectroscopy
TM	transverse magnetic
TNIL	thermal nanoimprint lithography
UV	ultraviolet
UV-NIL	ultraviolet-nanoimprint lithography

VBG	volume Bragg grating
VCSEL	vertical-cavity surface-emitting laser

Publications

Journal publications

- **X. Liu**, S. Lebedkin, H. Besser, W. Pfleging, S. Prinz, M. Wissmann, P. Schwab, I. Nazarenko, M. Guttmann, M. Kappes, and U. Lemmer, "Tailored surface-enhanced Raman nanopillar arrays fabricated by laser-assisted replication for biomolecular detection using organic semiconductor lasers," *ACS Nano*, vol. 9, pp. 260–270, 2015.
- **X. Liu**, S. Prinz, H. Besser, W. Pfleging, M. Wissmann, C. Vannahme, M. Guttmann, T. Mappes, S. Köber, C. Koos, and U. Lemmer, "Organic semiconductor distributed feedback laser pixels for lab-on-a-chip applications fabricated by laser-assisted replication," *Faraday Discussions*, vol. 174, pp. 153–164, 2014.
- **X. Liu**, P. Stefanou, B. Wang, T. Woggon, T. Mappes, U. Lemmer, "Organic semiconductor distributed feedback (DFB) laser as excitation source in Raman spectroscopy," *Optics Express*, vol. 21, pp. 28941–28947, 2013.
- **X. Liu**, S. Klinkhammer, Z. Wang, T. Wienhold, C. Vannahme, P.-J. Jakobs, A. Bacher, A. Muslija, T. Mappes, U. Lemmer, "Pump spot size dependent lasing threshold in organic semiconductor DFB lasers fabricated via nanograting transfer," *Optics Express*, vol. 21, pp. 27697–27706, 2013.
- **X. Liu**, S. Klinkhammer, K. Sudau, N. Mechau, C. Vannahme, J. Kaschke, T. Mappes, M. Wegener, and U. Lemmer, "Ink-jet-printed organic semiconductor distributed feedback laser," *Applied Physics Express*, vol. 5, pp. 072101, 2012.

- S. Schauer, **X. Liu**, M. Worgull, U. Lemmer, and H. Hölscher, "Shape-memory polymers as flexible resonator substrates for continuously tunable organic DFB lasers," *Optical Materials Express*, vol. 5, pp. 576–584, 2015.

- P. Brenner, L.-M. Fleig, **X. Liu**, A. Welle, S. Bräse, and U. Lemmer, "Degradation mechanisms of polyfluorene-based organic semiconductor lasers under ambient and oxygen-free conditions," *Journal of Polymer Science Part B: Polymer Physics*, DOI: 10.1002/polb.23733, 2015.

- F. Nickel, C. Sprau, M. F. G. Klein, P. Kapetana, N. Christ, **X. Liu**, S. Klinkhammer, U. Lemmer, and A. Colsmann, "Spatial mapping of photocurrents in organic solar cells comprising wedge-shaped absorber layers for an efficient material screening," *Solar Energy Materials and Solar Cells*, vol. 104, pp. 18–22, 2012.

- S. Klinkhammer, **X. Liu**, K. Huska, Y. Shen, S. Vanderheiden, S. Valouch, C. Vannahme, S. Bräse, T. Mappes, and U. Lemmer, "Continuously tunable solution-processed organic semiconductor DFB lasers pumped by laser diode," *Optics Express*, vol. 20, pp. 6357–6364, 2012.

SPIE Proceedings

- **X. Liu**, S. Lebedkin, T. Mappes, S. Köber, C. Koos, M. Kappes, and U. Lemmer, "Organic semiconductor distributed feedback laser as excitation source in Raman spectroscopy using free-beam and fibre coupling," *Proc. SPIE* 9137, 91370Y (2014) (**Best Student Paper Award**).

Conferences (personal contributions)

- **X. Liu**, S. Lebedkin, W. Pfleging, H. Besser, M. Wissmann, I. Nazarenko, T. Mappes, S. Köber, C. Koos, M. Kappes, and U. Lemmer, "Surface-enhanced Raman spectroscopy (SERS) using nanopillar arrays as functional substrates and an organic semiconductor DFB laser as excitation source," CLEO, San Jose, 8.-13. June 2014 (oral presentation).
- **X. Liu**, S. Lebedkin, T. Mappes, S. Köber, C. Koos, M. Kappes, and U. Lemmer, "Organic semiconductor distributed feedback laser as excitation source in Raman spectroscopy using free-beam and fibre coupling," SPIE Photonics Europe, Brussels, 14.-17. April 2014 (poster).
- **X. Liu**, S. Klinkhammer, Z. Wang, K. Sudau, N. Mechau, C. Vannahme, T. Mappes, and U. Lemmer, "Organic semiconductor distributed feedback (DFB) laser pixels fabricated via nanograting transfer and ink-jet printing," CLEO Europe, Munich, 12.-16. May 2013 (oral presentation).
- **X. Liu**, S. Klinkhammer, K. Huska, T. Mappes, U. Lemmer, "Continuously tunable solution-processed organic distributed feedback (DFB) lasers by horizontal dipping," EOS Annual Meeting, Aberdeen, 25.-28. September 2012 (oral presentation).
- **X. Liu**, S. Klinkhammer, U. Lemmer, "Liquid-processed continuously tunable organic DFB-lasers," Winter School of Organic Electronics, Heidelberg, 9.-12. December 2010 (poster).

Supervised Theses

- Panagiotis Stefanou, Organische Halbleiterlaser als Anregungsquelle für Raman-Streuung, Diplomarbeit, 2013.
- Stephan Prinz, Herstellung von funktionellen Nanostrukturen durch laserunterstütztes Heißprägen, Master thesis, 2013.
- Dong Zhao, Laser modes between two organic DFB lasers, KSOP research project, 2012.
- Matthias Sauer, Laserdiodenpumpen für kommerzielle organische Halbleiterlaser, Diplomarbeit, 2012.
- Florian Krauß, Lebensdauer von organischen Materialien mit optischer Verstärkung, Bachelor thesis, 2011.

Acknowledgments

I would like to thank all the people who have helped me in one way or another during my work at KIT.

First I would like to express my sincere gratitude to my advisor, Prof. Dr. Uli Lemmer, who sparked my interest in optoelectronics and provided precious support and encouragement at key times. He was always there with new suggestions and intriguing discussions, yet also had patience and open mind for my naive ideas. His ongoing confidence in me was so essential to the completion of this work.

Next I would like to thank Prof. María A. Díaz-García from University of Alicante for her friendly interest in this thesis and for agreeing to be my second examiner.

My special thanks go to:

Priv.-Doz. Dr. Timo Mappes for accepting me in his Young Investigator Network (YIN) and evoking basic ideas for Photonics4Life project. His enthusiasm and timely response to my questions supported me in times of doubt, even after his new career beginning at Carl Zeiss AG.

Dr. Sebastian Köber as my research group Leader for two years. Many thanks for his considerate supervision and valuable suggestions.

Dr. Sergei Lebedkin for his generous cooperation at Institute of Nanotechnology (INT). He helped me to construct my SERS-setup at INT from scratch, assisted me to solve almost all the problems during experiments and encouraged me with his expert knowledge and calmness. Without his support, it would not have been possible for me to complete the SERS measurements and the whole work.

Dr. Christoph Vannahme for sharing his enormous knowledge and experience on fabrication of all-polymer lab-on-chip systems. He is the most trustful colleague, coauthor and cooperator. I could not remember how many times he helped me deposit the anti-adhesion layer on the silicon master stamps. His patience and enthusiasm always make me happy to work with him. And I am also thankful for his introduction about Indiaca, which always reminds me of the traditional Chinese game Jianzi.

Dr. Sönke Klinkhammer for giving me encouragement since my master thesis. My work is based on his great ideas and achievements.

Heino Besser and Dr. Wilhelm Pfleging for their ongoing support on laser-assisted replication.

Klaus Huska for his generous and valuable advices on the establishment of horizontal dipping and the measurement setup for Raman experiments.

Patrick Schwab for his FDTD simulations for SERS nanopillar arrays and fruitful discussions about LSPR.

Kai Sudau and Dr. Norman Mechau for their support on ink-jet-printed organic semiconductor laser.

The colleagues from Visolas GmbH: Dr. Thomas Woggon, Dr. Johannes Barth and Matthias Sauer for attending the OLaser meeting and contributing great ideas as industry partner.

I also would like to thank all of the LTI scientific staffs for our collaboration and enjoyable social events. My current and former office mates, Anne Habermehl, Carsten Eschenbaum, Nico Bolse, Florian Maier-Flaig, Lena-Maria Fleig, Tobias Bocksrocker, Jan Brückner, have given me a very comfortable and happy time in Room 219. Amos Egel helped me in understanding the waveguide theory and getting familiar with LightTools. Siegfried Kettlitz and André Gall organized the great LTI Movie-Night. And thanks a lot for the tasty desserts and fruitful discussions on the weekly MNOS meeting, for warm Christmas party and for our memorable outing events.

And I would like to express my gratitude to other LTI colleagues:

Mario Sütsch, Hans Vögele and Klaus Ochs for their timely mechanical support. Whenever I met problems with equipments in the lab, they are the first aids to rely on. They always discussed the technical draft drawings with me patiently and tried their best to finish every single mechanical part of consistent high quality. Christian Kayser and Thorsten Feldmann for their ongoing clean room support. Felix Geislhöringer for warm helps with his electrical expertise. Astrid Dittrich and Claudia Holeisen for their helps on many administrative issues. Zhenhao Zhang and Min Zhang for their consistent encouragement in mother tongue.

I give my sincerest gratitude to my IMT-colleagues, who have helped me in the past few years: Tobias Wienhold and Uwe Bog for their continuous support and encouragement; Markus Wissmann and Dr. Markus Guttmann for collaboration in KNMF project on laser-assisted replication; Senta Schauer and Dr. Hendrik Hölcher for collaboration in lasers based on shape memory polymer; Richard Thelen and Tobias Meier for AFM introduction; Dr. Uwe Köhler and Heike Fornasier for clean room introduction, Peter-Jürgen Jakobs, Andreas Bacher and Marie-Kristin Nees for Ebeam lithography; Alban Muslija for reactive-ion etching; Paul Abaffy for SEM measurements and Au/Ag sputtering.

I shall not forget to express my appreciation to all of "my" students Panagiotis Stefanou, Stephan Prinz, Matthias Sauer, Florian Krauss, Bohui Wang, Dong Zhao for their efforts and contribution to my thesis.

It is my pleasure to thank Karlsruhe School of Optics and Photonics (KSOP) for recruiting me in her organization and offering me the fellowship with other intelligent research staffs. Ninette Stürzl has suggested me to do SERS researches at INT. Johannes Kaschke helped me deposit the protection layer on the polymer chips before ink-jet printing. Christoph Krämmer and Sarah Wiegele shared with me their experience on Raman spectroscopy and organic laser dyes.

Financial support from the Carl-Zeiss-Stiftung scholarship through my research is gratefully acknowledged.

I owe a great debt to Christoph Vannahme, Timo Mappes, Sebastian Köber, Philipp Brenner, Anne Habermehl and Nico Bolse for their detailed comments on my dissertation at every stage.

Finally, my deepest gratitude goes to friends and my families. Particularly to my parents, who always have confidence in me and are willing to share my joys and pains. And to Hongyan, for always standing by my side and making my life complete.